MARINE RESOURCE DAMAGE ASSESSMENT

Marine Resource Damage Assessment

Liability and Compensation for Environmental Damage

Edited by

FRANK MAES

Department of Public International Law & Maritime Institute,
University of Ghent,
Belgium

Springer

A C.I.P. Catalogue record for this book is available from the Library of Congress.

ISBN-13 978-90-481-6846-0 (PB)
ISBN-13 978-1-4020-3368-1 (e-book)

Published by Springer,
P.O. Box 17, 3300 AA Dordrecht, The Netherlands.

www.springeronline.com

Cover images:
Sinking of the 'Prestige' off the coast of Spain, 19th of November in 2002. The pictures are used with the permission of Douane Française – François Guyader.

Printed on acid-free paper

Printed in the Netherlands.

TABLE OF CONTENTS

PART IV. TRADE-OFF BETWEEN ENVIRONMENTAL AND SOCIO-ECONOMIC FACTORS IN THE BELGIAN PART OF THE NORTH SEA

INTRODUCTION

Prof. Dr. FRANK MAES

Department of Public International Law & Maritime Institute
Ghent University, Belgium

Universiteitstraat 6, B – Ghent,
Tel.: + 32 9 264 68 95, Fax: + 32 9 264 69 89
E-mail: Frank.Maes@UGent.be

The contributions to this book are partly the output of a research project conducted between 1998 and 2002 in Belgium and a conference held at the Ghent University, Belgium, in June 2003 on the topic "*Marine Resource Damage Assessment and Compensation for Environmental Damage*" (MARE-DASM). The Belgian Federal Science Policy sponsored both the research and the conference.

MARE-DASM research focused on: (i) the estimation and distribution of marine contaminants in order to assess their long term effects (ecotoxicology); (ii) the integration of these result into a Biological Effects SubModel and a mathematical model assessing the risks associated with accidental spillage of oil at sea and the damage this can cause (modelling); (iii) the assessment of the willingness to pay for ecological damage, based on the Contingent Valuation Method (economics); (iv) the development and evaluation of measures to be taken in order to guarantee a sustainable use of the Belgian part of the North Sea, taking into account the economic and social interests and values (social economics); (v) the potential to develop technical and legal procedures that allow ecological damage to the marine environment to be evaluated and compensated, taking into account constraints in national and international liability legislation (legal). The objective of the conference was to compare these research results with other research, expertise and state practice related to: (i) the assessment of long term effects of hazardous substances released into the marine environment by means of models; (ii) the use of economic evaluation tools for ecological damage to the marine environment; and (iii) the potentials and constraints to recover environmental/ecological damage through liability regimes and compensation funds.

F. Maes (ed.), Marine Resource Damage Assessment, Liability and Compensation for Environmental Damage, xv–xvii.

This research project and the conference were multi-disciplinary in nature, as reflected in the contributions to this book. The main focus is on civil liability regimes to compensate for environmental damage, the economic valuation of damage to the marine environment and the application of the Contingent Valuation Method in Belgium. Finally, an assessment is made of the economic value of user functions in the Belgian part of the North Sea and an existing mathematical model for estimating oil pollution damage at sea has been refined.

According to principle 13 of the 1992 Rio Declaration: "*States shall develop national law regarding liability and compensation for the victims of pollution and other environmental damage. States shall also co-operate in an expeditious and more determined manner to further develop international law regarding liability and compensation for adverse effects of environmental damage...*".

Within the EU a long discussion took place concerning a draft Directive on the prevention and the restoration of significant environmental damage with the aim of harmonizing environmental liability law within the EU. In April 2004 this draft was approved by the European Parliament and the Council and became Directive 2004/35/EC on environmental liability with regard to the prevention and remedying of environmental damage, narrowing down the application of the initial draft. E. BRANS explores the content of this Directive and its relation to other EC Directives, focusing on compensation of restoration measures. The topic of compensation or restoration for environmental damage is also the focus of other authors in this book. L. MEDIN reports on the Latvian laws that seem to allow compensation for ecological damage and explains how this can be calculated according to Latvian legislation. Reference is made to one case that is still pending. L. PROOT and H. BOCKEN report on the potentials and problems for compensating environmental damage in Belgian legislation. In particular, the Law on the protection of the marine environment in the areas under jurisdiction of Belgium (1999) *prima facie* seems to leave an opportunity for this kind of compensation. However, it seems that the EU Directive and this Belgian law do not intent to divert from the international conventions on civil liability in case of oil pollution by ships. J. NICHOLS, as Deputy Director of the 1992 IOPC Fund, explains how the parties to the 1992 Civil Liability Convention and the 1992 Fund Convention have tried to find a universally acceptable solution for the compensation for environmental damage caused by oil pollution from ships. Recent accidents with oil tankers (Erika in 1999 and the Prestige in 2002) have stimulated the discussion on compensation for environmental damage within the EU and the International Maritime Organization. G. GONSAELES reports on these developments from a EU perspective and the impact of EC decision making on the international regime for oil pollution damage. In first instance the EU intended to act unilaterally with a proposal for a COPE Fund, which finally found its way to the universal level. This so-called third tier of compensation was approved in May 2003 at an International Conference on the Establishment of a Supplementary Fund for Oil Pollution Damage under the auspices of the IMO. The idea of an additional fund is not new, since a comparable fund has already been in existence for a long time in Canada. Experiences with the Canadian Ship-source Oil Pollution Fund (SSOPFund) are explained in the contribution of K. MACINNIS. MACINNIS is Administrator of this Fund and admits that even in Canada the opportunities of this fund have not yet been fully explored in the courts.

Compensation for economic loss of profits is internationally accepted and proving this loss *prima facie* looks easier compared to environmental damage, as shown in the contribution of F. DOUVERE. The discussion on compensation for environmental damage, however, cannot be held without input from colleagues with expertise in theoretical economics. Ultimately, compensation for environmental damage has to be expressed in monetary terms. P. NUNES and A. DE BLAEIJ, in their contribution on the use of non-market valuation methods, explain the variety of solutions proposed by experts in economics. One of those methods, the Contingent Valuation Method (CVM), has been applied in Belgium in the frame of the MARE-DASM research. K. VAN BIERVLIET, D. LE ROY and P. NUNES report on the methodology used and the results achieved in the performance of a CVM applied to a theoretical accidental oil spill along the Belgian coast. Finally, the MARE-DASM research explored the idea of using mathematical models for estimating oil pollution damage at sea. S. SCORY has worked out this idea in his contribution to this book by using a case study corresponding to a hypothetical scenario based on recent accidents that took place along the Belgian coast in the period 2001-2003.

Acknowledgements

The editor would like to acknowledge the assistance of Gwendoline GONSAELES in the preparation of this book. Without her help this book would have not been camera ready in due time. The editor would further like to express his acknowledgment to the publishers for their willing to take a risk in publishing contributions of authors representing different scientific disciplines. Although multi-disciplinary research is very much promoted, it is not common for a publisher to follow this track. Finally, this research would not have been possible without the financial support of the Belgian Federal Science Policy in the frame of the Programme 'Sustainable Management of the North Sea' (1997-2002).

PART I.

LEGAL ASPECTS OF ENVIRONMENTAL DAMAGE

CHAPTER 1

ESTIMATING DAMAGES UNDER THE 2004 EC DIRECTIVE ON ENVIRONMENTAL LIABILITY

EDWARD H.P. BRANS

Pels Rijcken & Droogleever Fortuijn

P.O. Box 11756, 2502 AT The Hague, The Netherlands
Tel.: + 31 70 515 32 32 - Fax: + 31 70 515 30 66
E-mail: ehp.brans@prdf.nl

1. INTRODUCTION

In April 2004, the European Parliament and the Council adopted Directive 2004/35/EC on environmental liability with regard to the prevention and remedying of environmental damage[1]. The Directive is another step in the development of the EU's nature protection policy. It complements existing ex ante EC nature conservation regimes such as the regimes established by the Wild Birds and Habitats Directives[2]. In that respect it is also not surprising that the new Directive seeks a connection with these Directives and uses some of their definitions and starting points.

---

[*] This paper is an update and revision of an article published in *Env. Liability* 2002, 135-147. A revised version of this paper is published in Environments Law Review, 2005, No. 4.

[1] Directive 2004/35/EC, *OJ* L 143/56, 21 April 2004. Member States have until the end of April 2007 to transpose this Directive into domestic law.
[2] Resp. Directive 79/409/EEG, *OJ* L 103/1, 1979 and Directive 92/43/EEG, *OJ* L 206/7, 1992.

F. Maes (ed.), Marine Resource Damage Assessment, Liability and Compensation for Environmental Damage, 3–25.

The focus of this article is on the Directive's measure of damages and its rules for assessing damages. As will be shown, under the Environment Liability Directive damages are preferably assessed on the basis of the actual costs of restoration measures. The Directive therefore contains guidelines on how to determine the level and nature of the restoration measures to be taken to bring back the injured natural resources to baseline condition and to compensate for the interim loss in natural resource services. These rules thus determine the extent of the responsible party's liability and can therefore be considered rules for assessing damages.

The Directive's measure of damages and its guidelines on natural resource damage assessment are partly based on US laws and their damage assessment regulations. For that reason, comparisons will regularly be made with the US federal laws that allow certain governmental agencies to act as trustee and claim and assess damages for injuries to what could be called public natural resources. I shall also look at international civil liability conventions that cover damage to natural resources. According to Article 4 of the new Directive, the Directive will not apply if international conventions regulate liability for the type of incident that occurred. Consequently, it is interesting to explore what type of damage is recoverable under these conventions and whether or not there exists a difference between the new EC liability regime and these international liability conventions.

Before addressing the above subjects, an overview is provided of the Environmental Liability Directive[3]. First, a general overview is given of the Directive's key features, then a more in-depth analysis is provided regarding the natural resources covered, and the issue of standing.

2. THE SCOPE OF THE REGIME - GENERAL OVERVIEW

The Directive imposes a strict or fault-based liability - depending on the type of activity involved - on the operator of an occupational activity for damage to protected species and natural habitats, contamination of land and damage to waters covered by the Water Framework Directive[4] (provided the damage is above a certain threshold) (Article 3 and Article 2(1)). Operators who undertake an activity that is covered by the EC legislation listed in Annex III of the Directive can be held strictly liable for the above three types of harm (for which the overarching term 'environmental damage' is used). The EC legislation listed in Annex III includes Directive 96/61/EC concerning integrated pollution prevention and control (IPPC Directive) and legislation on the transportation of dangerous substances, on waste management operations and on the direct release of genetically modified organisms into the environment. Most of the activities covered by the listed EC legislation can be considered environmentally risky activities. A fault-based liability is imposed on

---

[3] For a more detailed overview the reader is referred to FOGLEMAN, V., The Environmental Liability Directive, *Env. Liability*, 101-115 (2004) and KRÄMER, L., Directive 2004/35/CE on Environmental Liability, in: BETLEM, G. & BRANS, E.H.P. (ed.), *Environmental Liability in the EU*, London (forthcoming).

[4] Directive 2000/60/EC establishing a framework for Community action in the field of water policy, *OJ* L327/1, 2000.

operators of non-listed occupational activities. These operators can only be held liable for damage to protected species and natural habitats and not for the other types of harm mentioned.

There are several situations that are exempted from the Directive. For example, environmental damage that arises from an incident in respect of which liability or compensation falls within the scope of a number of listed international civil liability conventions is not covered by the Directive.[5] Nor does the Directive apply to environmental damage caused by an emission or incident that took place before 30 April 2007, the date Member States should have transposed the Directive into domestic law[6]. Moreover, an operator cannot be held liable under the Directive for environmental damage caused by an act of armed conflict, hostilities or civil war, or by a natural phenomenon of exceptional, inevitable and irresistible character[7]. In addition, an operator may escape liability if he proves that the damage was caused by a third party, provided appropriate safety measures were in place, or if he proves that the damage resulted from compliance with an order or instruction from a public authority[8]. The Directive also allows Member States the discretion to exempt an operator from liability where the operator demonstrates that he was not at fault or negligent and that the environmental damage caused resulted from an emission or event expressly authorized by the regulatory authority[9]. Apart from the so-called regulatory compliance defence, Member States may also decide to exempt an operator from liability where the operator demonstrates that he was not at fault or negligent and that the environmental damage caused resulted from an emission or event not considered likely to cause environmental damage according to the state of scientific and technical knowledge at the time the emission was released or the activity took place[10].

Liability under the Directive is not limited to a certain ceiling. This, however, does not mean that liability is unlimited. As noted earlier, the Directive contains a set of guidelines on selecting the most appropriate measures to remedy the environmental damage caused. These guidelines have been introduced, among other things, to prevent the liable operator from being confronted with disproportionate costly restoration measures or a disproportionate claim (see below). According to these guidelines only reasonable restoration measures are to be taken to remedy the environmental damage caused, thereby taking into account - among other things - the costs of implementing the various restoration options[11].

---

[5] See Article 4(2-4) and Annex IV of the Directive.
[6] See Article 17 of the Directive.
[7] See Article 4(1) of the Directive.
[8] See Article 8(3) of the Directive.
[9] See Article 8(4)(a) of the Directive.
[10] See Article 8(4)(b) of the Directive.
[11] See para. 1.3.1 of Annex II.

3. WHAT NATURAL RESOURCES ARE COVERED?

As noted earlier, the Environmental Liability Directive imposes a strict or fault-based liability - depending on the type of activity involved - for damage to protected species and habitats, for contamination of land and for damage to waters covered by the Water Framework Directive. Operators who undertake an activity listed in Annex III of the Directive can be held strictly liable for these three types of harm. Operators of non-listed occupational activities can only be held liable for damage to protected species and habitats and not for damage to the waters regulated by the Water Framework Directive or for the contamination of land.

3.1. Damage to protected species and habitats

Earlier versions of the Environmental Liability Directive employed the term 'biodiversity damage' to point out what natural resources are covered by the EC regime[12]. The term is, however, not part of the final Directive. One of the reasons for deleting the term is that it gives rise to confusion because the interpretation of the term 'biodiversity' in these earlier versions of the Directive differed from more authoritative and generally accepted interpretations, such as that provided by the 1992 Convention on Biological Diversity[13]. In this Convention biodiversity is defined as *the number, variety and variability of all species of plants, animals and micro-organisms as well as the ecosystems of which they are part* (Article 2). Biodiversity is thus more than just the number of species in a certain area[14]. In fact, there are four levels at which biodiversity is assessed: genetic diversity within a species, the variability among species, functional diversity which refers to the variety of biological functions of ecosystems, and ecosystem diversity which refers to the variety of communities of organisms within particular habitats[15].
In these earlier versions of the Directive the term 'biodiversity' was defined in a more limited way. For instance, according to Article 2(1) of a 2002 draft of the Directive 'biodiversity' means:

> *"natural habitats and species listed in Annex I to Directive 79/409/EEC, or in Annexes I, II and IV to Directive 92/43/EEC10, or habitats and species, which are not covered by the those Directives, for which areas of*

---

[12] See in this respect the 2000 White Paper on Environmental Liability (COM(2000) 66 final) and the EC's 2001 Working Paper on Prevention and Restoration of Significant Environmental Damage, both available at http://www.europe.eu.int/comm./environment/liability.

[13] (1992) 31 *ILM* 818.

[14] HARPER, J.L. & HAWKSWORTH, D.L., Preface in: HAWKSWORTH, D.L. (ed.)(1996), *Biodiversity. Measurement and Estimation,* London, 7-10.

[15] HARPER, J.L. & HAWKSWORTH, D.L., *ib.*, 6; UNEP (1995), *Global Biodiversity Assessment*, Cambridge, 27 et seq; THORNE-MILLER, B. (1999), *The Living Ocean. Understanding and Protecting Biodiversity*, Washington D.C., 6-7; NUNES, P.A.L.D., VAN DEN BERGH, J.C.J.M. and NIJKAMP, P. (2000), *Ecological - Economic Analysis and Valuation of Biodiversity,* Amsterdam/Rotterdam, 3-5.

> *protection or conservation have been designated pursuant to the relevant Member States' legislation on nature conservation"*[16].

The focus was thus on natural resources which are (already) protected under EC nature conservation Directives, which are considered to be of public interest and whose conservation status requires the designation of special protection areas (the Natura 2000 network) and not on biodiversity as defined under the 1992 Convention on Biological Diversity[17]. Since the final Directive uses a comparable starting point with regard to its scope, it has been decided not to use the term "biodiversity" any more[18]. Consequently, damage to biodiversity as defined in the 1992 Biodiversity Convention is not covered by the new regime (which, however, does not necessarily mean that loss of biodiversity cannot be considered when planning restoration measures to bring back the impacted natural resources to baseline condition).

So what natural resources are covered by the Directive on Environmental Liability? The scope of the new regime is limited to the species and natural habitats protected by the Wild Birds and Habitats Directives[19]. In addition, Member States have, under certain conditions, the option to include species and natural habitats not covered by the Wild Birds and Habitats Directives. One of the conditons is that these natural resources are protected by national protection and conservation laws. In view of this, it would appear that not all natural resources are covered by the Directive.
In earlier versions of the Directive the regime was limited to natural resources located in the protection areas that have to be designated by the Member States under the Wild Birds and Habitats Directives. The protection areas that are (going to be) designated form together a European ecological network (called Natura 2000). It is expected that finally about 10% to 12% of the territory of the European Union will be classified as a Natura 2000 site[20]. Under the earlier versions of the Directive damage to natural resources located outside the Natura 2000 sites was thus not covered, even where the species and habitats concerned are listed in the annexes to the Wild Birds and Habitats directives.
The geographical limitation of the regime to Natura 2000 sites was considered by NGO's and others as a serious restriction to the scope of the regime[21] In response to the opposition of NGO's and others against this limitation, it was finally decided to

---

[16] COM(2002) 17 final.

[17] For further details on the designation process, the consequences of the delay in designating sites, relevant case law, etc., see: SCOTT, J. (1998), *EC Environmental Law*, Harlow, 111 *et seq.*; BACKES, Ch., Nature Conservation, *YEEL*, 331-338 (2001); BETLEM, G. & BRANS, E.H.P., The Future Role of Civil Liability for Environmental Damage in the EU, *YEEL*, 194-195 (2002); BRANS, E.H.P. (2001), *Liability for Damage to Public Natural Resources. Standing and Damage Assessment*, The Hague/london/New York, 195-199. See also the documents available at http://www.europe.eu.int/comm/environment/nature/ home.htm.

[18] The term is used in the preamble of the Directive, not in the Directive itself.

[19] See Article 2(3) of the Directive.

[20] See further on the designation process, http://www.europa.eu.int/comm/environment/nature.

[21] See further BETLEM, G. & BRANS, E.H.P., *o.c.*, 192-193.

set aside the idea of limiting the liability regime to Natura 2000 sites. Now the regime covers all natural resources protected by the Wild Birds and Habitats Directives, wherever located (and the natural resources that are included by Member States).

It should be noted that damage to the abovementioned category of natural resources is only recoverable if a certain threshold is met. According to Article 2(1) of the Directive this will be the case if the damage is of such a nature that it has '*significant adverse effects on reaching or maintaining the favourable conservation status*' of the habitats and species concerned. In order to determine whether a particular incident has such an effect, it is noted in the Directive that this needs to be determined by comparing the condition the natural resources are in after the incident occurred with the baseline condition of these natural resources and the associated natural resource services[22]. Baseline condition is the condition the impacted natural resources and services would have had if the incident that caused the damage had not taken place[23]. In Annex I of the Directive guidance is provided on how to determine whether a certain event caused significant adverse changes to the impacted natural resources and the services provided by these natural resources. One of the factors to be taken into account is the capacity of these natural resources for natural recovery.

It should be noted that the above threshold criteria are based on the Habitats Directive. In that respect one may expect that, for instance, case law of the European Court of Justice regarding the interpretation of the relevant clauses of the Habitats Directive, is also relevant for the Environmental Liability Directive[24].

3.2. Damage to waters covered by the Water Framework Directive

The Directive also covers damage to the waters covered by the Water Framework Directive[25]. Under this Directive, Member States are required to designate protection areas in river basins in order to protect surface water and groundwater and to conserve habitats and species directly depending on the waters. No precise figures are available, but it seems that by including the above waters the geographical boundaries of the liability regime are expanded considerably.

According to the Directive damage to the above areas and waters is only recoverable if the damage is of such a nature that it '*adversely affects the ecological status, ecological potential and/or chemical status*' of these waters (Article 2(18)). The Water Framework Directive itself provides criteria for determining whether or not the ecological status or potential of the waters have been adversely affected.

---

[22] *Ib.*

[23] See Article 2(1)(a), (13) and (14) of the Directive.

[24] An interesting case in this respect is the ECJ Case C-127/02 and especially the conclusion of the Advocate General J. KOKOTT.

[25] Directive 2000/60/EC establishing a framework for Community action in the field of water policy, *OJ* L327/1, 2000.

It should be noted that the Directive excludes operators of non-listed activities from liability for damage to the waters covered by the Water Framework Directive. There is no explanation why this category of operators is excluded from liability.

3.3. Soil pollution

The Directive also covers soil pollution damage. However, this type of damage is only recoverable if it '*creates a significant risk of human health being adversely affected*' (Article 2(1)(c)) and if it is caused by a listed potentially dangerous activity. If the latter is not the case, the operator will escape liability (at least under EC law).

The Directive or preamble does not explain why an operator is exempted from liability if the soil pollution damage is caused by a non-listed activity. The decision is understandable for cases where damage has been caused to nature areas not covered by the future regime. However, if damage is caused to nature sites falling under the scope of the Directive, in my view it is inefficient to separate soil pollution damage caused by listed activities from that caused by non-listed activities. In most cases, whatever the cause of the incident, it will be necessary first to take measures to clean up the polluted area before actions can be taken to enhance the restoration process and help the injured natural resources and services to return to baseline condition.

Another point of concern is the criteria included in the Directive for determining whether the soil pollution is of such an extent that it exceeds the damage threshold. These criteria are not very detailed and precise. There is no reference to concentration limits of polluting substances or other objective criteria to determine whether or not clean-up is necessary[26]. Furthermore, the effects of the contamination on the environment seem not to be decisive when deciding on the necessity of clean-up measures. Clear clean-up objectives are also missing. In short, this part of the Directive does not seem to add much (at least not for the Member States that have detailed legislation in this area, such the Netherlands and Belgium). It is not argued that the Directive should not contain a provision on liability for soil pollution. However, in my view it would have been more efficient if the Directive would have allowed recovering costs of clean-up measures that are to be taken in order to be able to take measures that will enhance the restoration process and help returning the injured natural resources and services to baseline condition (see below).

[26] Some information on the goal of clean-up measures and on how to determine whether or not the contamination is such that is poses a significant risk of adversely affecting human health is provided in paragraph 2 of Annex II of the Directive.

3.4. Owned and unowned natural resources

An interesting aspect of the Directive is that it covers damage to natural resources regardless the legal status of the natural resources. If the natural resources impacted are covered by this Directive and the damage is above a certain threshold (see Article 2(1)(a-c)), then the damage is in principle recoverable under the Directive. No distinction is made between owned or unowned natural resources. From an environmental point of view, this is an improvement.
The current state of tort law means that various obstacles exist that prevent a full recovery of damages for injury to natural resources[27]. First, the injury to unowned natural resources does not easily fit into the categories of compensable types of harm; it is not a pecuniary loss, nor can it be qualified as a recoverable non-pecuniary or immaterial loss[28]. Furthermore, since no one suffers a direct loss if damage is caused to unowned natural resources, the question arises of who may claim for injury to these natural resources. Another problem concerns owned natural resources. Although, the issue of standing does not pose a major difficulty here, the fact that natural resources are owned does not guarantee that the natural resources, including the environmentally significant natural resources, are going to be restored. The owner might be unwilling to initiate an action for damages or to spend the monetary compensation received on restoration.
The Environmental Liability Directive solves these problems by establishing a sort of trusteeship construction. Under the Directive, the public authorities have the right to require operators responsible for causing damage to the natural resource covered by the new regime, to take the necessary restorative measures[29]. If operators fail to comply with the request, the public authorities may take the measures themselves and recover the costs[30]. The legal status of the natural resources concerned is immaterial.

It should be noted that the directive does not provide for a recovery of compensation for 'traditional damage' (personal injury, damage to goods and property, and pure economic loss). These heads of damage are excluded from the scope of the directive because these types of loss are already recoverable under the liability laws of the Member States[31]. It should be noted, however, that given the above there is one exception to this rule. Damage to real property is recoverable under the Directive, but only if the property contains a natural habitat covered by the Directive. However, since such a habitat directly or indirectly serves the public interests and is

---

[27] Cf. COM(2000) 66 final, p. 12; CARETTE, A. (1997), *Herstel van en vergoeding voor aantasting aan niet-toegeëigende milieubestanddelen*, Antwerpen; LARSSON, M.-L. (1999), *The Law of Environmental Damage*, London; BRANS, E.H.P. (2001), *o.c.*, 241 *et seq.*
[28] BRANS, E.H.P. (2001), *o.c.*, 28 *et seq.*
[29] See Article 6(2)(c) of the Directive.
[30] See Article 6(3) and Article 8 of the Directive. The competent authority is not under a duty to take restoration measures if the operator refuses to take the measures. Initially it was proposed to impose such a duty on the Member States, but this proved not to be acceptable to the Member States.
[31] Earlier versions of the Directive included a proposal to impose liability for these types of harm. See for instance the White Paper on Environmental Liability, COM(2000) 66 final.

therefore of public value, under the Directive it is not the property owner who has a right to file a claim for damages but the public authorities. They act as a sort of trustee for the natural resources covered by the regime. Since there is, as was noted earlier, no duty for any property owner to file a claim or to use recovered sums for restoration purposes, and damage to these public natural resources may therefore not be restored, this construction was introduced to tackle this problem.

4. STANDING - THE STATE AS A TRUSTEE FOR PUBLIC NATURAL RESOURCES

According to Article 6 of the Directive, public authorities have the right to require the operator who caused a significant damage to the natural resources covered to take the necessary restoration measures. The authority to order restoration concerns all natural resources falling under the scope of the regime and includes owned as well as unowned natural resources (which is understandable from the point of view that the natural resources concerned are of public value).[32] The Directive thus empowers the public authorities to act as a sort of trustee for the natural resources concerned. In that respect the EC regime is comparable to the US federal laws that provide a legal basis for claiming compensation for injury to public natural resources (*i.e.* CERCLA[33] and the 1990 Oil Pollution Act[34] (OPA)). Under these laws, particular public authorities act on behalf of the public as a trustee for certain natural resources and are empowered to assess and claim damages for injuries to these natural resources[35].

If the operator fails to comply with the 'request', the competent authority may take the restoration measures themselves but can recover the costs of the measures[36]. The competent authority is not under a duty to remedy the damage caused if the polluter refuses to take the necessary restoration measures, cannot be identified, or is exempted from liability[37].
In previous versions of the Directive, Member States were required to take the necessary restoration measures if the operator refuses to act or cannot be identified. However, the proposed "subsidiary responsibility" for Member States caused a considerable debate among the Member States[38]. The Member States objected to the proposal because of the costs involved. It was suggested to set up some kind of fund

---

[32] The decision not to distinguish between owned and unowned natural resources is a logical decision as such a distinction would make the regime highly impractical. It should, however, be recognized that structuring a regime as is being done might have legal and economic consequences for landowners. See further BRANS, E.H.P. (2001), *o.c.*, 26-28.
[33] 42 U.S.C. §§ 9601 *et seq.*
[34] 33 U.S.C. §§ 2701 *et seq.*
[35] OPA 33 U.S.C. §§ 2701(20); 2702(b), 2706(a); CERCLA 42 U.S.C. §§ 9601(16), 9607(a)(4). For further details, see BRANS, E.H.P. (2001), *o.c.*, 65 *et seq.*
[36] See Article 6(3) jo Article 8 of the Directive.
[37] See Article 6 (3) of the Directive.
[38] Minutes of the 2439th Council Meeting, Environmental, Luxemburg, 25 June 2002, 10013/02 (Press 180).

to finance the restoration measures, but this proposal was not acceptable either. So currently Member States are not under an obligation to take restoration measures if the responsible party refuses to take such measures, cannot be identified, or is exempted from liability.

4.1. Public interest groups

Compared to the Commission's White Paper on Environmental Liability (COM(2000) 66 def.) the position of public interest groups has changed fundamentally. Unlike the White Paper, NGOs are now excluded from the right to claim compensation from alleged polluters. Instead, qualified NGOs are afforded the right to request the competent authorities that action be taken against a polluter (Article 12). Provided the request for action and the accompanying information and data show in a 'sufficiently plausible' manner that environmental damage has been caused, the public authorities are under a duty to consider the requests of NGOs and have to inform the NGOs of its decision(s) (Article 12(4)). The public interest groups have the right to bring legal proceedings for review of the public authorities' response to their requests for action (Article 13). The judicial review procedure is directed at a review of the '*procedural and substantial legality of the decisions, acts or failure to act*' of the competent authority. Although not specifically mentioned, the right of review also seems to cover the legality of decisions of the competent authority regarding the nature and extent of the restoration measures and maybe also the terms of a negotiated settlement with the responsible party. The right of review does not exist if no request for action was submitted.
Apart from the right to request the competent authority to take action and the right to start review procedures, public interests groups have the right to submit observations regarding the restoration measures to taken. Given the text of Article 7(4), (2), it seems that the right to submit observations is limited to the issue of which of the selected restoration measures shall be implemented and not on issues such as the type of injury determination studies to be undertaken or the development of possible restoration alternatives[39]. No provision is included in the Directive regarding the notification of (draft) restoration plans and the option of submitting observations.

For some time now, the EC has been preparing the ratification of the 1998 Aarhus Convention on Access to Information, Public Participation in Decision-making and Access to Justice in Environmental Matters[40]. Many Member States have also ratified the convention or are preparing ratification[41]. Given the provisions of the EC

---

[39] In the US too the public is given the opportunity to comment on a proposed restoration plan. However, the public has also the option under certain conditions to comment on a settlement agreement before it is approved in court. 61 Fed. Reg. 441, 442 (1996). See further BRANS, E.H.P. (2001), *o.c.*, 91-92 & 233-234.

[40] The text of the convention is avalable at http://unece.org/env/pp. For an overview of EC activities taking place to allow ratification of the convention by the EC, see: http://www.europa.eu.int/comm/environment/aarhus/index.htm.

[41] See http://unece.org/env/pp/ctreaty.htm.

Environmental Liability Directive on access to justice for public interest groups, the question arises whether these provisions are in line with this convention. Overall this seems to be the case, but there is one important exception.

Under Article 9(3) of the Aarhus Convention NGO's have the right to not only challenge acts or omissions by public authorities which contravene national environmental law, but also acts or omissions of private persons that do so. According to the implementation guide to the Convention, Article 9(3) has been introduced to give citizens standing to go to court or other review body, to enforce environmental law[42]. So the idea behind this provision is to give public interest groups (and others) the option of participating in the enforcement process, even where it concerns the acts or omissions of private persons. Now, what is striking is that if one compares the Directive's provisions on access to justice for public interest groups with the Aarhus Convention, there is a difference between the two. Under the Directive, NGO's have no right to challenge the acts or omissions of a private person. They may request the public authorities to take action and start a review procedure against the competent authority only if, according to the public interest group, no adequate action is taken. Given the fact that the EC is preparing ratification of the Aarhus Convention, the provisions in the Directive on access to justice is surprising. Maybe it is thought that because it is stipulated in the Directive that the Directive shall not prevent Member States from maintaining or adopting more stringent provisions in relation to the prevention and remedying of environmental damage[43], Member States themselves will take appropriate action to prevent a violation of the Aarhus Convention.

4.2. Property owners

As noted earlier, if damage is caused to natural resources that are subject to property rights, such as a part of a Natura 2000 area, it is the owner who has the right to sue. However, under civil law there is no duty to file a claim or to use the compensation obtained for restoration purposes. In order to ensure the restoration of damaged natural resources that are of Community interest, it was decided to introduce deviating rules on standing. Under these rules, the public authorities are empowered to require polluters to take appropriate restoration measures or to take the measures themselves and recover the costs even if the natural resources concerned are owned. Property owners have not been awarded direct access to justice. Instead, they have the right under the Directive to request the public authorities that action be taken against a polluter, provided they suffer environmental damage (Article 12(1)). And, if restorative measures are going to be taken on their land or waters, these property owners are given the option to submit observations with regard to the nature of the measures (Article 7(4)).

The Directive does not prevent a person whose property has been affected by a certain incident from filing a claim against the polluter on the basis of national law.

[42] ECE (2000), *The Aarhus Convention. An Implementation Guide*, New York/Geneva, 130.

[43] See Article 16 of the Directive.

However, the proposed Directive prohibits double recovery of the damage (Article 16(2)). No rules have been included in the proposal on how to prevent overlapping claims and draw a line between public and private losses[44]. From US case law it appears that this is not an easy subject and that courts may experience difficulties in deciding on the issue[45].

5. DAMAGE ASSESSMENT

5.1. Measure of damages

One of the primary objectives of the Directive is to restore damage caused to these natural resources. The Directive therefore emphasizes restoration and chooses restoration costs as the primary and preferred method to assess damages[46]. However, because it takes time to restore the natural resources to baseline condition, that is, the condition of the natural resources before the injury occurred, the polluter will also be held liable for the loss or impairment of natural resources and natural resource services during the restoration period (interim losses)[47]. In addition to restoration costs (and interim losses), the responsible party can be held liable for the costs of assessing damages as well as the administrative, legal and enforcement costs, the costs of data collection and monitoring and oversight costs[48].

The Directive's measure of damages is comparable with the measure of damages of the US Oil Pollution Act. That is not that surpising since the European Commission used the 1990 Oil Pollution Act and its natural resource damage assessment regulations (NRDA rules)[49] as a source of inspiration. Under this US law a responsible party may be held liable for *e.g.* (i) the cost of restoring the injured natural resources and services to baseline condition; (ii) the cost of restoration that compensates for the interim loss of resources and services that occur from the time of the incident untill recovery of such resources and services to baseline condition; and (iii) the reasonable cost of assessing damages[50].

So both measures of damages have similarities[51]. However, there are differences. For instance, unlike the EC Directive the Oil Pollution Act also provides for a number of private claims[52]. Apart from that, it is my impression that the Oil Pollution Act and its NRDA rules places more emphasis on the loss of human use services during the

---

[44] The US natural resource damage assessment regulations do provide some guidance on how to distinguish private from public losses. See 61 Fed. Reg. 445, 459, 484-85 (1996) and BRANS, E.H.P. (2001), *o.c.*, 87-90.

[45] See further BRANS, E.H.P. (2001), *o.c.*, 87-88 (with references).

[46] See Article 7(1) and Annex II of COM(2002) 17.

[47] See Article 2(11),(13) and Annex II, para. 1(c) and (d).

[48] See Article 8(2) jo 2(16) of the Directive.

[49] These NRDA rules are published in 61 Fed.Reg. 1996, p. 440 *et seq.* See also the guidance documents that further explain these NRDA rules. Available at http://www.darp.noaa.gov.

[50] 33 U.S.C.A § 2702(b)(2)(A). See also 61 Fed.Reg. 1996, p. 441 *et seq.*

[51] See also BURLINGTON, L., Valuing Natural Resource Damages: A Translantic Lesson, in: BETLEM, G. & BRANS, E.H.P., *o.c.*

[52] An example of such a claim is the claim for loss of profits or loss of earning capacity due to the injury of natural resources. See 33 U.S.C.A § 2702(b)(2)(E).

period of restoration than the EC Directive. Examples of such losses are the loss of recreational beach use or recreational fishing. The loss of natural resource services, which is defined in the Directive as the loss of '*functions performed by a natural resource for the benefits of another natural resource or the public*'[53], is an undeniable part of the Directive's measure of damages[54], but to me it seems that there is less emphasis on the loss of human use services during the period that action is taken to return the injured natural resources and services to baseline condition. It is, however, to be noted that the guidelines of Annex II of the Directive are less detailed than the NRDA rules of the Oil Pollution Act[55].

5.2. Primary and compensatory restoration measures

The injury, loss or destruction of natural resources often has significant side effects. The damage to natural resources may result for instance in a loss or impairment of natural resources services during the period of (natural) recovery. Natural resources services are the functions a natural resource may have for the benefit of other natural resources and for the public. A coastal wetland or river basin, for instance, provides food and nesting habitat for birds and other species, clean water for fish populations, and is important for biodiversity maintenance and for pollution assimilation. Examples of human benefits deriving from natural resources such as a coastal wetland or a river basin include recreational fishing and boating, beach use, wildlife viewing, hiking, waterway navigation and subsistence hunting[56]. So if due to releases of hazardous substances an injury is caused to a coastal wetland or a river basin then this may affect ecological or inter-resource services and human uses or services.

According to the Directive, damages are to be assessed on the basis of measures taken to restore the injured natural resources and/or impaired natural resources services to baseline condition. The Directive uses the term 'primary restoration' measures for this type of measures[57]. The term is also used in the NRDA rules to the Oil Pollution Act.

The Directive requires that the public authority that is awarded the task of assessing the natural resource damage and determining what primary restoration measures are going to be taken, considers natural recovery - in which no action is taken to directly restore the injured natural resources and services - as an appropriate primary restoration action[58]. It is also permitted to re-create elsewhere a replacement habitat

---

[53] See Article 2(13) of the Directive and para. 1(c) and d) of Annex II.

[54] See Article 2(11), (13) of the Directive and para. 1(c) and d) of Annex II.

[55] It is beyond the scope of this contribution, but given para. 1.2.3 of Annex II the loss of human uses during the period of recovery may play a role when determining the scale of the restoration measures to compensate for the interim loss of natural resources and services. As will be shown, the Directive uses the term 'compensatory restoration' measures for this type of measures. See para. 1(c) of Annex II.

[56] For an overview of services provided by natural resources, see DE GROOT, R.S. (1992), *Functions of Nature*, Amsterdam, 15 *et seq.*

[57] See Annex II, para. 1(a).

[58] Cf. Annex II, para. 3.1.1.

or to take other off-side measures[59]. The Directive uses the term 'complementary restoration' measures for such measures. The NRDA rules to the Oil Pollution Act do not use this term. However, the taking of off-side restoration measures is also allowed under these NRDA rules. It is, however, considered to be a primary restoration measure.

Because primary (or complementary) restoration measures do not compensate for the loss of ecological and human services during the restoration period, 'compensatory restoration' measures are to be taken to compensate for that loss (Article 2(11) and para. 1(d) of Annex II). To give an example, if an oil spill or other incident results in a significant damage of a number of acres of wetland and natural recovery is the most appropriate option here, then during the recovery period (some) wetland services are lost or impaired. Under the Directive, the polluter has to compensate for this temporary loss of wetland services. This could be done through the provision of additional services of the same type as those lost, for instance by creating additional acres of wetland or enhancing degraded wetland[60]. These new acres of wetland are expected to provide (during its functional life-span) services equal to those lost from the period of the spill until full recovery of the affected wetland[61]. The cost of creating or enhancing the acres of wetland is the amount of compensation to be paid for the loss of services during the period of recovery, i.e., the interim losses.

5.2.1. Methods to scale compensatory restoration measures[62]

This paragraph deals with the question of how to determine the appropriate scale of the compensatory restoration measures. The Directive provides some guidance on the issue[63]. However, because the guidelines are not very detailed I also look at the NRDA rules that accompany CERCLA and the US Oil Pollution Act and that have been established to guide the US trustees in assessing damages for natural resource injuries and the development and selection of restoration options. As noted earlier, the US approach regarding the assessment of damages for these injuries is comparable to that of the Directive. In fact, this part of the Directive is based on the Oil Pollution Act and its NRDA rules. The terminology

---

[59] See Annex II, para. 1.1.2. The purpose of the so-called 'complementary restoration' measures is to provide a similar level of natural resources/and or services as would have been provided if the damaged site had been returned to baseline condition. It is noted in Annex II, para. 1.1.2, that where possible the alternative site should be linked to the damaged site.

[60] According to para. 1.1.3 of Annex II, compensatory measures consists of additional improvements to protected natural habitats and species or waters at either the damaged site or at an alternative site.

[61] JONES, C.A. & PEASE, K.A., Restoration-Based Compensation Measures in Natural Resource Liability Statutes, 15 *Cont. Econ. Pol.* 112 (1997); MAZZOTTA, M.J., OPALUCH, J.J. and GRIGALUNAS, T.A., Natural Resource Damage Assessments: The Role of Resource Restoration, 34 *Nat. Res. J.*, 170-173. (1994); JULIUS, B.E. & OSBORN, T. (1992), *Wetland Compensation Analysis. Greenhill Oil Spill*, NOAA, Silver Spring MD, 2.

[62] For a more detailed discussion there is referred to the NRDA-rules (61 Fed. Reg. 440), the documents available at http://www.darp.noaa.gov and BRANS, E.H.P. (2001), *o.c.*, 131-154.

[63] See para. 3.1.3-3.1.8 of Annex II.

used in the Directives's guidance document is also almost similar to the US NRDA rules, and so is the structure of the proposed damage assessment procedure.

Scaling. The type and scale of the compensatory restoration measures depends on the level and speed of recovery of the injured natural resources and services under the primary (or complementary) restoration measures. For example, if a primary (or complementary) restoration action accelerates the natural recovery of the impacted natural resources, the interim losses are likely to be less than if no such measures are taken[64]. To determine the scale of the compensatory restoration measures, one needs to determine the benefits of a certain primary (or complementary) restoration action. Only after one has determined the extent and duration of service losses under one or more specific primary (or complementary) restoration actions, is it possible to determine what scale of compensatory restoration is necessary to compensate for interim losses. Scaling is thus required to ensure that the identified restoration alternatives appropriately compensate for the interim losses[65].
Scaling involves adjusting the size of the compensatory restoration action to ensure that the gains provided by this action equal the interim loss in services due to the incident and over the period of restoration[66]. There are two basic approaches to scaling compensatory restoration measures: the service-to-service approach, or the valuation approach.

The service-to-service approach. The service-to-service approach is used to determine the scale of a compensatory restoration action that provides natural resources and services of the same type and quality as, and of comparable value, to those lost due to the incident[67]. This approach is used if there is an equivalence between the resources and services lost and the resources and services gained through compensatory restoration. Habitat Equivalency Analysis (HEA) is one method that can be used here. The principal concept underlying this method is that '*the public can be compensated for interim losses of habitat resources through habitat replacement projects providing additional resources of the same type and quality*'[68].

[64] An illustrative figure is included in JONES, C.A. & PEASE, K.A., *l.c.*, 114 and BRANS, E.H.P. (2001), *o.c.*, p. 132.
[65] The OPA NRDA rule indicates that scaling also applies to primary restoration actions that involve replacement and acquisition of equivalent resources. 15 CFR 990.53(d); 61 Fed. Reg. 452-453 (1996). It is unclear whether the Commission considers this necessary.
[66] Because the duration of the injury differs from the lifespan of the replacement action, equivalency is calculated in terms of the present discounted value of service flows lost due to resource injuries and service gains from the compensatory restoration action. CHAPMAN, D. et al. (1998) *Calculating Resource Compensation: An Application of the Service-to-Service Approach to the Blackbird Mine Hazardous Waste Site*, Techn. Paper No. 97-1, Silver Spring MD, 2. For more details, see: NOAA (1999), *Discounting and the Treatment of Uncertainty in Natural Resource Damage Assessments* Tech. Paper No. 99-1, Silver Spring MD (http://www.darp.noaa.gov, 10 August 2004).
[67] 15 CFR § 990.53(d)(2).
[68] NOAA (1995 rev. 1999), *Habitat Equivalency Analysis: An Overview*, Silver Spring MD, 1 (http://www.darp.noaa.gov, 13 October 2004).

The following example is illustrative (in general terms only): if due to an incident acres of wetland become seriously oiled and for cost-effectiveness reasons no primary restoration measures can be taken, then during the period of natural recovery there is a loss of natural resource services. Presume that planting salt marsh vegetation at a location close to the affected area is an appropriate compensatory restoration action because it is a cost-effective means of replacing resources and resource services substantially similar to those lost. The planted salt marsh vegetation is expected to provide habitat for bird species and a variety of aquatic species, some of which are recreationally and commercially important. HEA can be used to determine the size of the compensatory measures, such that the total services provided by the compensatory habitat over its functional lifespan is equal to the total services lost due to the injury (from the period of the incident until recovery of the injured area has occurred)[69]. In order to determine the scale of the compensatory restoration project, account is taken of, among other things, the type and level of services lost, the trajectory of recovery of the impacted wetland back to baseline condition, the benefits of a primary restoration project (if undertaken) and the productivity of the created habitat. The costs of the planting of acres of salt marsh vegetation to compensate for the interim loss of natural resource services, is the measure of damages.
HEA is now frequently used to determine the scale of compensatory restoration projects. In fact, it is estimated that it has been used in 50-80 percent of the recent federal has been cases involving natural resource injuries[70].

The valuation approach. In cases where it is inappropriate to use the service-to-service approach, or where trustees are unable to produce restoration alternatives that provide natural resources or services of the same type, quality, or comparable value as those injured, then trustees may use the valuation approach[71]. There are two options here: value-to-value and value-to-cost scaling[72].
With value-to-value scaling, trustees calculate the value of the injured natural resources and services and look for compensatory restoration actions that generate benefits with a value equal to those lost. The responsible party is liable for the cost of implementing the compensatory restoration project that is expected to generate the equivalent value, but not for the calculated monetary value of the interim losses[73]. The value of the gains of a restoration project and the value of interim losses is assessed with help of economic valuation techniques, including the travel

---

[69] JULIUS, B.E. (1997), *US v. Melvin A Fisher*, NOAA, Silver Spring MD, 2-3 (damage assessment report/testimony). Because the losses and gains are occurring at different times, a discount rate is applied to translate all of the terms to what they are worth at the present year. Discounting is a crucial component of HEA and can be compared with the notion of interest. See for further details: NOAA (1999), *Discounting and the Treatment of Uncertainty in Natural Resource Damage Assessments*, Tech. Paper No. 99-1, Silver Spring MD (http://www.darp.noaa.gov, 10 August 2004).

[70] See TOMASI, T., KEALY, M.J. & ROCKEL, M., Scaling Compensatory Restoration under the 1990 Oil Pollution Ac' in *Proc. IOSC*, 247 (1999)(CD Rom).

[71] 61 Fed. Reg. 442, 452-453 (1996).

[72] *Ib.*, 453.

[73] *Ib.*, 442.

cost method, hedonic pricing, conjoint analysis and contingent valuation (the latter, however, is seldom used, if at all)[74].

Value-to-cost scaling may be used if the valuation of natural resources and services provided by a compensatory restoration action cannot be performed within a reasonable time frame or at a reasonable cost. In that case, trustees estimate the dollar value of the injured resources and lost services - with help of the earlier mentioned economic valuation techniques - and select a compensatory restoration action based on that dollar amount[75]. Value-to-cost scaling is the more 'traditional' way of assessing damages for injury to natural resources.

Outlook. The Directive's framework for assessing damages - as well as the US NRDA rules - focuses on developing and implementing plans to restore damaged natural resources and services, rather than assessing the dollar value of the damage to such resources and services. The use of economic valuation methods is not ruled out by the Directive, but is now mainly used to 'determine what scale of compensatory restoration provides an equivalent value to the lost services' (to use a US quote)[76]. The responsible party pays for the cost of implementing the compensatory restoration project, not the monetary value of interim losses (unless the value-to-cost approach is used). One of the advantages is that it makes use of less controversial economic valuation techniques (and therefore more acceptable to the potentially liable parties).

The damage assessment methodology described above has proven to be successful in the U.S and has been approved by the courts[77]. In that respect the Directive's choice for the US approach is understandable. However, the Directive's guidelines for assessing damages are less detailed than the US NRDA rules. Because assessing damages for natural resource injuries remains a highly complex task, consideration should be given to further developing Annex II to the Directive.

5.2.2. Selecting the preferred restoration alternative

According to the Directive's damage assessment guidelines, and similar to the US NRDA rules, a reasonable range of restoration alternatives has to be developed, each consisting of a primary and compensatory component[78]. The competent authority

---

[74] For further information on these methods and the conditions under which these can be used (including the controversial Contingent Valuation Method), see: KOLSTAD, C.D. (2000), *Environmental* Economics, Oxford; PERMAN, R. et al. (1999), *Natural Resource & Environmental Economics*, Essex; 58 Fed. Reg. 4601 (1993); BRANS, E.H.P. (2001), *o.c.*, 103-108, 154-162. See also *infra* in this book PROOT, L. & BOCKEN, H., *Environmental damge and Belgian law* (Part I – Chapter 2 – 2.3.2.) and NUNES, P.A.L.D. & DE BLAEIJ, A.T., *Economic assessment of marine quality benefits: applying the use of non-market valuation methods* (Part III - Chapter 7) and VAN BIERVLIET, K., LE ROY D. & NUNES, P.A.L.D., *A contingent valuation study of an accidental oil spill along the Belgian coast* (Part III – Chapter 8).

[75] *Ib.*, 453.

[76] 61 Fed. Reg. 442 (1996).

[77] Cf. General Electric Co. v. NOAA, 128 F.3d 767 (D.C. Cir. 1997).

[78] Annex II, para. 3.2.1.

of the Member State then has to select the most appropriate restoration alternative on the basis of a set of selection criteria (Annex II, para. 3.2.1).
The available restoration options have to be evaluated on the bases of criteria such as the technical feasibility of the measures, the likelihood of success, the extent to which each alternative is expected to return the injured resources and services to baseline condition and to compensate for interim losses, and the cost of the restoration alternatives[79]. If two or more restoration alternatives are equally preferable, the public authorities have to select the most cost-effective restoration alternative[80]. There is no hierarchy of selection criteria and the selection of restoration actions is incident-specific.
The whole process of identifying, evaluating and selecting restoration alternatives is also important for determining when restoration costs become grossly disproportionate. The Directive does not set a specific standard (or numerical ratio) for determining at which point the cost of restoration becomes grossly disproportionate to the monetary value of the natural resources injured or to the benefits of a particular restoration alternative. Instead, it was decided to require the national competent authorities to weigh all of the aforesaid (and other) criteria when selecting the most appropriate restoration alternative(s)[81].
In the US the same set of criteria is used not only to select appropriate restoration options, but also to determine whether a certain restoration option is cost-reasonable. In the US it is expected that if all listed factors are considered, this will constitute a quantitative cost-benefit analysis and ensure that a cost-effective alternative is selected[82]. It is thought that this approach will protect the responsible party '*against selection of an inappropriate costly alternative*' and make certain that the preferred actions are '*commensurate with the value of the natural resource losses*'[83].

5.3. Designating competent authorities

Assessing the nature and extent of the natural resources injuries and identifying a reasonable range of appropriate restoration alternatives, each comprising a primary and compensatory restoration component, are complex tasks. For that reason, the Directive requires the designation of a 'competent authority' and gives them the task of assessing the significance of the natural resource injuries and of determining appropriate restoration plans (Article 11(2)). To guide these competent authorities in conducting these tasks, the guidelines of Annex II were developed.
The decision to designate a competent authority and to delegate the above tasks and to provide guidance on the assessment of damages, might prove to be crucial for the success of the Directive. In the US, statutes enacted prior to CERCLA and OPA did not designate a specific agency that was entitled to recover damages and that was responsible for the recovery and restoration of the injured natural resources: '*[T]he lack*

[79] *Ib.*
[80] Annex II, para. 3.2.2.
[81] See para 1.3.1 of Annex II.
[82] 61 Fed. Reg. 490 (1996).
[83] 61 Fed.Reg. 454, 490 (1996).

of specific delegated responsibility meant that no agency had the motivation or mandate to develop the structured programme needed to perform [...] assessments on a regular basis'[84]. Under CERCLA and OPA, specific governmental agencies have been designated to act as trustee for certain natural resources and are empowered to assess injury and damages, to present claims to recover damages and are responsible for the spending of recovered sums. The impression is that this, together with the availability of detailed NRDA guidelines, contributed to a rise in NRD claims in the US. Relevant too is the establishment of regional damage assessment teams, which support the trustees in assessing natural resource injuries, the development and evaluation of restoration plans and the recovery of damages through negotiation or litigation[85]. These teams - composed of biologists, resource economists and lawyers - conduct natural resource damage assessments on a regular basis and therefore have a lot of experience in determining the nature and extent of the natural resource injuries, the development and evaluation of restoration alternatives and the implementation of restoration strategies. For the success of the Directive, the EC should consider setting up comparable teams or maybe to give the European Environmental Agency the task of supporting the national 'competent authorities' in performing assessments and the development and evaluation of restoration alternatives.

5.4. Cooperative assessments

An important element of the US NRDA rules is the promotion of cooperative assessments[86]. Not so long ago most natural resource damage assessments were litigation-driven. This often led to a damage assessment process where both the responsible party and the trustees conducted separate assessment studies; a time consuming and costly approach. Furthermore, conducting separate studies in the context of litigation proved to influence the quality of the studies.
The 1996 NRDA rules replace the litigation driven model by an approach more directed at cooperation. The trustees are now under a requirement to give the responsible party the opportunity to participate in the damage assessment[87]. However, the degree of participation varies by incident and is to be determined on an incident-specific basis. The responsible party and the trustees may undertake joint assessment activities and develop restoration plans in close cooperation, but the responsible party's activities may also be limited to the opportunity to comment on the trustees' documents and plans[88].

---

[84] BRIGHTON, W.D. & ASKMAN, D.F., The Role of Government Trustees in Recovering Compensation for Injury to Natural Resources in WETTERSTEIN, P. (ed.)(1997), *Harm to the Environment*, Oxford, 182.
[85] For further information, see: http://www.darp.noaa.gov/.
[86] 61 Fed. Reg. (1996) 443.
[87] *Ib.* See also 15 CFR § 990.14(c).
[88] 61 Fed. Reg. (1996) at 443-4.

The approach has proven to be successful and is now applied on a regular basis[89]. Positive effects of the approach are cost savings, a shortened time to restoration and a reduced potential for litigation[90]. However, the latter is also due to procedures in CERCLA and OPA that encourage settlements[91].
The European Commission supported the concept of cooperative assessments[92]. However, nothing on cooperative assessments the Directive contains nor is guidance provided on the nature and degree of a possible cooperation, neither is made clear under what circumstances or requirements a responsible party may participate in a damage assessment. The OPA's regulations provide guidance on when to invite a responsible party, how to determine the level of participation, the conclusion of formal agreements, the funding of assessment studies, the documentation of decisions during the assessment process and the involvement of the public in the whole assessment process. In my view, it is regrettable that the Directive does not promote cooperative assessments. It probably would have made the regime more efficient and cheaper.

6. INTERNATIONAL CONVENTIONS

According to Article 4(2) and (4) of the Directive, the Directive will not apply to environmental damage arising from an incident in respect of which liability or compensation falls witin the scope of a number of listed international civil liability conventions[93]. The listed conventions include the 1992 International Convention on Civil Liability for Oil Pollution Damage, the 1992 International Convention on the Establishment of an International Fund for Compensation for Oil Pollution (hereafter the 1992 Oil Pollution Conventions), the 2001 Bunker Fuel Oil Pollution Damage Convention and the 1997 Convention on Supplementary Compensation is that for Nuclear Damage. A number of the listed conventions have not yet entered into force, which means that the Directive for the moment will apply to incidents covered by these conventions[94].

---

[89] For further details, see: HELTON, D. (2000), *The Benefits of Cooperative Natural Resource Damage Assessments*, Silver Spring, http://www.darp.noaa.gov and BRANS, E.H.P. (2001), *o.c.*, 145-147. See further http://www.darp.noaa.gov for documents regarding cooperative assessments.
[90] HELTON, D. (2000), *Ib.*
[91] BRANS, E.H.P. (2001), *o.c.*, 232-234.
[92] See COM(2002) 17, p. 3 and Annex II, para. 3.2.4.
[93] These conventions are listed in Annex IV and V.
[94] Examples of conventions which are listed but have not entered into force are the 2001 Bunker Fuel Oil Pollution Damage Convention and the 1989 Convention on on Civil Liability for Damage Caused durring Carriage of Dangerous Goods by Road, Rail and Inland Navigation Vessels.
See further on the relationship of the Directive with International conventions: FAURE, F. & WANG, H., Liability for oil pollution – the EU approach, *Env. Liability*, 64-66 (2004).

Most international civil liability conventions that cover environmental damage apply the same or almost the same damage definition: damage to the environment is recoverable but the 'compensation for impairment of the environment [is] limited to the costs of reasonable measures of reinstatement actually undertaken or to be undertaken'[95]. The definition was developed in 1984 and was part of the 1984 Protocol that was drafted to amend the 1969 International Convention on Civil Liability for Oil Pollution Damage and the associated 1971 Fund Convention (and never entered into force)[96]. The pollution damage definition was finally incorporated in the 1992 Protocols that amend the above oil pollution conventions[97].
Probably because these oil pollution conventions have been successfully tested over the years, many other international conventions that cover environmental harm apply the same or a comparable damage definition. Examples are the 1989 CRTD convention, the 1996 HNS Convention, the 1999 Protocol to the 1989 Basel Convention on the Control and Transboundary Movement of Hazardous Wastes and their Disposal, and the 2001 Bunker Fuel Oil Pollution Damage Convention[98].
A problem with the 1984 damage definition is that damage to natural resources is only recoverable if measures are taken to reinstate the impaired environment and the costs of such measures prove to be reasonable[99]. If it is physically impossible to restore the damaged environment, if no one is willing to take restoration measures, or if the costs involved are disproportionate, the polluter cannot be held liable for the damage done and will escape liability. Nor does the damage definition seem to allow the taking of – to use the terminology of the EC Directive – complementary restoration measures if it proves impossible to take – to use the terminology of the EC Directive – primary restoration measures and to fully restore the impacted natural resources.
An interesting development is now that some of the more recently drafted conventions do allow the taking of complementary restoration measures. These conventions use the above damage definition, but added a provision to it which makes it possible under certain conditions to introduce the equivalent of the damaged or destroyed natural resources if it is impossible for technical or financial

---

[95] For an overview of international civil liability conventions and their damage definitions, see e.g. DE LA FAYETTE, L., The Concept of Environmental Damage in International Liability Regimes, in BOWMAN, M., & BOYLE, A. (2002), *Environmental Damage in International and Comparative Law*, Oxford, 181 *et seq.* and BRANS, E.H.P. (2001), *o.c.*, 311 *et seq.* See also *infra* in this book NICHOLS, J., *Scope of compensation for environmental damage under the 1992 Civil Liability Convention and the 1992 Fund Convention* (Part II – Chapter 4) and GONSAELES, G., *The impact of EC decision-making on the international regime for oil pollution damage* (Part II – Chapter 6).
[96] Resp. 9 *ILM* 45 (1970) and 11 *ILM* 284 (1972).
[97] International Transport Treaties, Suppl. 17, I-459/476 (Sept. 1993). See further on the oil pollution conventions, OOSTERVEEN, W., Liability for damage resulting from oil pollution. From the perspective of an EU member state, in BETLEM, G. & BRANS, E.H.P. (ed.), *o.c.*
[98] For further details, see: DE LA FAYETTE, L., *o.c.*, 181 *et seq.* and BRANS, E.H.P. (2001), *o.c.*, 366 *et seq.*
[99] For a more detailed analysis of shortcomings of these international conventions (and their damage definition), see: BRANS, E.H.P. (2001), *o.c.*, 318-404 and DE LA FAYETTE, L., *o.c.*, 181 *et seq.*

reasons to take primary restoration measures[100]. So these conventions do permit the taking of measures to, for instance, re-create elsewhere a habitat to compensate for the loss of or injury to a habitat in the impacted area (provided such measures are reasonable and appropriate). Interestingly, the policy of the 1992 International Oil Pollution Fund, which administers the 1992 Oil Pollution Conventions, has also recently been changed and now the costs of reinstatement measures '*taken at some distance from, but still within the general vicinity of, the damaged area*' are in principle compensable (provided it can be demonstrated that these measures actually enhance the recovery of the damaged components of the environment, and furthermore that the measures are taken or going to be taken and the costs involved are reasonable)[101]. So under some international liability conventions the costs of complementary restoration measures are recoverable.

Interim losses are clearly not recoverable under the international civil liability conventions. This is also the case with the more recently drafted conventions. Nor does the Claims Manual of the 1992 International Oil Pollution Fund refer to interim losses or compensatory restoration measures. Presumably the time it takes to restore the natural resources to baseline condition under the various available restoration options may be considered when selecting the appropriate restoration plan, but the conventions itself do not allow for a recovery of the costs of measures to compensate for interim losses (compensatory restoration measures).

Another difference with the EC Directive is that most of the international civil liability conventions do not provide any guidance on determining the reasonableness of restoration measures[102]. Nor is the goal of the restoration measures specified. It is unclear, for instance, whether lost human services are to be considered when selecting restoration alternatives.

So if one compares the above with the Directive, it is obvious that the EU regime applies a different measure of damages and also provides more guidance on how to assess injury and damages. The most striking difference, however, concerns the compensation of interim losses. Unlike the EC Directive, under the international liability conventions a polluter cannot be held liable for the loss of natural resources or natural resource services during the period of recovery. With regard to most of the international regimes, a polluter will even escape liability if for whatever reason no primary restoration measures can be taken. As noted earlier, not all the conventions listed in Annex IV and V of the Directive seem to allow the taking of complementary restoration measures if it appears impossible to take primary restoration measures.

---

[100] See the 1997 Protocol to the Vienna Convention on Nuclear Damage and the 2003 Protocol to the UN/ECE Convention on the Transboundary Effects of Industrial Accidents. See also the 1993 Lugano Convention on Civil Liability for Damage Resulting from Activities Dangerous to the Environment.

[101] See Claims Manual of the International Oil Pollution Compensation Fund, November 2002, 29. See also *infra* in this book NICHOLS, J., *Scope of compensation for environmental damage under the 1992 Civil Liability Convention and the 1992 Fund Convention* (Part II – Chapter 4).

[102] An exception is the international oil pollution conventions. In the 1992 Fund's Claims Manual criteria are listed to determine the reasonableness of restoration measures (see also the Annex to 92FUND/A.7/4, 15 June 2002). Some guidance is also included in the 1997 Protocol to the Vienna Convention on Nuclear Damage (Article 2(4). See further BRANS, E.H.P. (2001), *o.c.*, 351 *et seq.*

7. CONCLUSION

In this article the focus was on the measure of damages and the Directive's framework for assessing damages. Based on the above analysis one may conclude that with regard to these issues, the liability Directive is to some extent comparable with the US laws that empower US trustees to file claims and assess damages for natural resource injuries. Similar to these US laws there is a clear preference for assessing damages on the basis of the cost of restoration measures. Striking is also that contrary to, for instance, the liability regime as proposed in the 2000 White Paper on Environmental Liability, the issue of recovering interim losses is clearly addressed and that with regard to the assessment of damages for the interim loss of natural resources and services, there is a (partial) reliance on the US NRDA rules.

In addition, it was shown that the measure of damages of international civil liability conventions differs from the one included in the EC Directive, especially where it concerns the compensation of interim losses. Interim losses are not recoverable under these conventions. An interesting development, however, is that some of the conventions mentioned do allow reinstatement measures to be taken some distance from the affected area, therby allowing the taking of complementary restoration measures.

A somewhat negative aspect of the Directive is that the guidelines on how to measure the extent of the injuries to natural resources and the loss of natural resources services, and how to determine the appropriate scale of the restoration measures, are not that detailed (at least not when compared to the OPA's NRDA rules). This might prove to be a disadvantage for those applying the regime, such as the competent authorities of the Member States. They may find the Directive difficult to apply, especially the part regarding the determination of the extent and nature of the natural resource injuries and the level and nature of the restoration measures to be taken to compensate the loss. In addition, it might also well be that public interest groups and others, including courts, might find it more difficult to check whether the Member State acted in accordance with the directive.

CHAPTER 2

ENVIRONMENTAL DAMAGE AND BELGIAN LAW

Mr. LAURENT PROOT & Prof. Dr. HUBERT BOCKEN

Centre for Environmental Law – Ghent University (Belgium)

Universiteitsstraat 4, 9000 Ghent, Belgium
Tel.: + 32 9 234 29 16, + 32 9 264 69 25 – Fax: + 32 9 234 29 17, + 32 9 264 69 90
E-mail: Laurent.Proot@ldr.be, Hubert.Bocken@UGent.be

1. DEFINITION OF THE NOTION 'ECOLOGICAL DAMAGE'

Before reviewing the Belgian liability rules regarding damage to the marine environment, the notion of 'environmental damage' has to be clearly defined, for several definitions are used through the different regulatory texts regarding liability for environmental damage[1]. The notion of 'environmental damage' is very broad and contains all kinds of damage caused by the impairment of the environment.

Common Belgian liability law uses the criterion of the 'ownership status' to distinguish environmental damage 'sensu strictu' from 'ecological damage'. Damage to appropriated elements of nature is defined as environmental damage 'sensu strictu'. This concerns the classical individual damage caused by environmental pollution (health damage, physical damage, cleaning costs, restoration costs, etc.).

Damage to non-appropriated elements of nature (*res nullius, res communes*[2]) is regarded as ecological damage[3].

---

[1] For instance the notions 'environmental damage', 'ecological damage', 'pure ecological damage', 'impairment of nature', 'damage to natural resources' are used.

[2] *E.g.* air, wild animals, ecosystems, the sea, wild fish (not breaded or economically dealed), …

[3] Next to the notion 'ecological damage', also the notions 'pure ecological damage' or 'impairment of nature' are used. However, all these notions refer to damage to not appropriated elements of nature. For a comprehensive review of the importance of the definition of the notion 'environmental damage', see BRANS, E.H.P. (2001), *Liability for damage to public natural resources. Standing, damage and damage assessment*, The Hague/London/New York, Kluwer Law International, 9.

F. Maes (ed.), Marine Resource Damage Assessment, Liability and Compensation for Environmental Damage, 27–41.

The Act of 20 January 1999 on the protection of the marine environment in the maritime areas under the jurisdiction of Belgium[4] also uses the ownership criterion to distinguish the notions 'environmental damage' and 'environmental disruption. More specifically, Article 2 of the Act on the protection of the marine environment defines 'environmental damage' as "*any damage, loss or prejudice suffered by an indentifiable natural or legal person as a result of a degradation of the marine environment whatever its cause*". 'Environmental disruption' means "*a negative impact on the marine environment, in so far as no damage is involved*"[5] (*i.e.* not suffered by an identifiable natural or legal person).

Next to the ownership criterion, '*the importance of the natural element for the ecosystem or for the human well-being*' can be used to define ecological damage. Here, ecological damage means damage to the ecological value of natural elements, irrespective of the ownership status. Consequently, ecological damage could also possibly occur in case of impairment of appropriated natural elements with a special ecological value[6].

The notions 'environmental damage' and 'ecological damage' (impairment of nature) are especially used in the European law order. In the US, the notion 'damage to natural resources' is used. Natural resources are among others land, fish, wild animals, air, water, drinking water resources which are owned or governed by the US, a state, a local authority, a foreign authority or an Indian tribe. No distinction is made between appropriated and not appropriated goods.

In the context of this contribution and according to the definitions of the Act on the protection of the marine environment, ecological damage means damage to non-appropriated elements of nature (the marine environment)[7]. However, ecological damage also means damage to the natural elements themselves, to the ecological function of the natural elements, for the problems and limitations of the Belgian liability law regarding compensation for ecological damage will be the same in both senses.

---

[4] *Belgian Official Journal* 12 March 1999, Second Edition. Hereinafter referred to as '*The Act on the protection of the marine environment*'.
[5] Article 2 of the Act on the protection of the marine environment.
[6] *E.g.* damage to a very ecologically valuable tree in a private garden.
[7] *E.g.* damage to fish, eco-systems, sea fauna and flora, etc.

2. OVERVIEW OF BELGIAN LIABILITY LAW WITH REGARD TO ECOLOGICAL DAMAGE

2.1. Common Belgian liability law and ecological damage[8]

2.1.1. Fault liability

The basic rules on liability for environmental damage can be found in Articles 1382-1383 of the Civil Code, establishing the fault liability. According to Article 1382 of the Civil Code, one can be held liable for the damage caused by his fault. Fault refers to socially unacceptable behaviour. In the context of environmental damage, the violation of a statutory or regulatory provision is the most important fault criterion, for there are a lot of regulatory provisions governing industrial activities[9]. However, next to the violation of a statute or regulation, there could be a fault in case of breach of the general duty of care (negligence). In order to determine whether someone has acted negligently, one has to compare the behaviour of the defendant with the presumed normal conduct of a reasonable man. In case of a professional, his actions must have been in conformity with the state of the art in his profession.

Next to the fault, there has to be damage. We distinguish classical environmental damage (health damage, physical damage, economic losses, etc.) and ecological damage[10].

Finally, to entail fault liability there has to be a causal link between the fault and the damage. In the Belgian law order, the theory of equivalence of conditions is applied. This means that a causal link is accepted, if the actual damage would not have occurred without the fault. The fault must be a necessary condition for the damage. Despite the broadness of the aforementioned concept of causation, difficulties of proof can occur, as the burden of proof lies with the victim in the Belgian law order.

2.1.2. Strict liability

Next to the classical fault liability there are several strict liability rules that can be used in the context of compensation for environmental damage. Strict liability implies that one can be held liable for damage, caused by his actions, even when these actions do not concern a fault.

First of all, Article 1384 § 1 of the Civil Code holds the custodian of a defective object liable for the damage caused by the defect. An object will be considered defective if it shows an abnormal characteristic. This liability rule is important in

---

[8] For a comprehensive review of Belgian liability law regarding ecological damage, see BOCKEN, H. The compensation of ecological damage in Belgium, in WETTERSTEIN, P. (1997), *Harm to the environment: the right to compensation and the assessment of damages*, Clarendon Press, Oxford, 143–158.

[9] One can think of the Flemish Waste Decree of 2 June 1981 or the Flemish Decree on the environmental Permit of 28 June 1985.

[10] See *supra*, section 1.

the context of environmental damage, for it has already been used several times in cases of pollution due to a defective installation[11].

Furthermore, Article 544 of the Civil Code is a legal basis for (even non-negligent) liability for excessive nuisances for the neighbours. On the basis of the aforementioned Article, one will be held liable if activities carried out on his land result in an excessive nuisance for the neighbours. In the context of the marine environment this will be a less important basis of liability.

Finally, there are several specific federal, regional and international statutes, providing for strict liability for damage caused by specific accidents. In that respect, Article 85 of the Act of 24 December 1976 forms the legal basis for the government to recover from the owner of the polluting products the expenses of the interventions of the Civil Protection and Fire Brigades, which have taken clean-up measures after a pollution incident[12]. The Flemish Soil Sanitation Decree also contains a strict liability rule for the person who causes soil pollution through an emission of polluting products[13]. On the international level, Article III(1) of the International Convention of 27 November 1992 on Civil Liability for Oil Pollution Damage, as amended, foresees a strict liability for the ship owner[14].

2.1.3. Limitations of common Belgian liability law concerning ecological damage

Two major problems arise if compensation for ecological damage is claimed according to common Belgian liability rules: the right of standing in case of non-appropriated elements of nature and the (financial) compensation for ecological damage, which can not be restored in kind.

Articles 17 and 18 of the Belgian Judicial Code require that a plaintiff must have a personal interest, if he claims compensation for ecological damage. The plaintiff must have an individual right or use of the damaged natural goods. This means that compensation claims for ecological damage to non-appropriated natural goods are not possible under common Belgian liability law[15].

Furthermore, if compensation in kind of ecological damage is not possible or reasonable, the question arises as to whether it is possible or desirable under Belgian liability law to grant financial compensation for that kind of damage. In that respect,

---

[11] In that respect, a polluted river or polluted soil has already been considered as a defective object. See BOCKEN, H. (1997), *l.c.*, 145.

[12] In the context of marine pollution, this Article was modified by Article 78 of the Act of 20 January 1999 on the protection of the marine environment. If it concerns marine pollution or pollution coming from a ship, the costs have to be recovered from the person who has caused the pollution according to international law (the ship owner).

[13] Article 25 of the Flemish Soil Sanitation Decree (*Belgian Official Journal* 22 February 1995) and BOCKEN, H. (1997), *l.c.*, 146. This Article could be a legal basis for the recovery of clean-up costs in case of oil pollution on beaches or dunes (Flemish territory) from the person who has caused the emission.

[14] Article III(1) of the International Convention on Civil Liability for Oil Pollution Damage (27 November 1992) provides: "*Except as provided in paragraphs 2 and 3 of this Article, the owner of a ship at the time of an incident, or, where the incident consists of a series of occurrences, at the time of the first such occurrence, shall be liable for any pollution damage caused by the ship as a result of the incident.*"

[15] See BOCKEN, H. (1997), *l.c.*, 172.

the problem of the monetary valuation of ecological damage (damage to natural goods which do not have an economic value) also has to be considered.

2.2. The Act of 20 January 1999 on the protection of the marine environment in the maritime areas under the jurisdiction of Belgium: overview of the liability system

2.2.1. Strict liability

The Act on the protection of the marine environment has introduced a specific liability regime to protect the marine environment in the maritime areas. In that respect, it brought environmental disruption/ecological damage under the scope of the liability system of the marine environment. Article 37 § 1 of the Act contains the following strict liability rule: '*Any one who causes damage or environmental disruption affecting the maritime areas following an accident or a violation of the legislation in force shall be liable to repair the damage, even if he has not committed any fault.*' This means that a claimant only has to prove that the damage is caused by an action of the defendant. No fault has to be proven. Three causes of exoneration are foreseen. According to Article 37 § 2 of the Act "*the person causing damage or environmental disruption shall not be held liable under § 1, if he can prove that the damage or environmental disruption:*
1° is exclusively the result of war, civil war, terrorism or a natural phenomenon of an exceptional, unavoidable and irresistible nature, or,
2° was caused entirely by a deliberate act or omission by a third party with the intention of causing damage or environmental disruption, and in so far as the third party concerned is not a representative or agent of the person held liable; or
3° was caused entirely by carelessness or any other detrimental act committed by an authority responsible for maintaining the lights or other navigational aids in the performance of its duties".
Furthermore, the Act on the protection of the marine environment does not prejudice the right of a responsible party for a pollution act to limit its liability in the cases and under the conditions stipulated in the applicable laws[16]. More specifically, this means that the limitations of liability of *e.g.* the International Convention on Civil Liability for Oil Pollution Damage or the Bunker Oil Convention, not yet ratified, remain into force, even when the pollution incident falls within the scope of the Act on the protection of the marine environment.

2.2.2. Damage and environmental disruption: right of standing

The Act on the protection of the marine environment has solved the abovementioned problem of the right of standing in case of ecological damage (to not appropriated natural goods). According to the Articles 37 § 3 and 40 § 2 of the Act on the protection of the marine environment, the State has the right to repair environmental

[16] Article 37 § 4 of the Act on the protection of the marine environment.

disruption[17]. Third parties also have the right to take measures for the restoration of ecological damage[18].

2.2.3. Restoration in kind

The restoration principle, contained in the Act on the protection of the marine areas, prescribes the restoration of the damaged natural goods in their original condition as far as possible[19]. The replacement of damaged goods by equivalent components is also possible[20]. However, when restoration in kind is not possible or not reasonable[21], the question arises as to whether 'equivalent restoration' of the ecological damage would be possible in the context of the Act on the protection of the marine environment. In fact, this concerns the question as to whether a monetary compensation for non-economically quantifiable losses in the marine environment would be possible. The answer to this question is not clear[22].

2.2.4. Environmental fund

The relevant indemnity for the environmental disruption, has to be paid to the Environmental Fund referred to in the table annexed to the Organic Act of 27 November 1990 establishing budgetary funds[23]. Consequently, the monetary compensation of the ecological damage is surely used in favour of the environment.

2.3. Monetary compensation for ecological damage to the marine environment in Belgian liability law: possible? Desirable?

2.3.1. The Act of 20 January 1999 on the Protection of the marine environment in the maritime areas under the jurisdiction of Belgium: restoration principles

The Act on the protection of the marine environment has introduced the concept of 'ecological damage' in the liability system of the marine environment. The Act clearly prefers restoration in kind or the replacement of the marine environment by

---

[17] Article 37 § 3 of the Act on the protection of the marine environment: "*A natural or legal person having suffered damage has the right to repair this damage. The State has the right to the repair of environmental disruption*". Article 40 § 2 of the Act on the protection of the marine environment clearly states: "*In the event of environmental disruption, the relevant indemnity shall be claimed by the State, without prejudice to the right of the other persons referred to in Article 37 § 5, to demand compensation, if need be, of the costs that they have incurred.*"

[18] Article 37 § 5 of the Act on the protection of the marine environment: "*The costs of measures to repair damage or environmental disruption taken by other persons than the liable person for the damage or the environmental disruption with a view to restore elements of the marine environment or replace them with equivalent elements, shall be indemnified by the liable person for the damage or the environmental disruption, in so far as the costs of these measures are not unreasonable in the light of the results to be achieved for the protection of the marine environment.*"

[19] Article 4 § 6 of the Act on the protection of the marine environment.

[20] Article 35 § 7 of the Act on the protection of the marine environment.

[21] Restoration in kind will be unreasonable, if the restoration costs would not be proportional with the replacement value of the natural good, see CARETTE, A. (1997), *Herstel en vergoeding voor aantasting aan niet toegeëigende milieubestanddelen*, Antwerpen, Intersentia, 166.

[22] See *infra*, section 2.3.

[23] Article 40 § 3 of the Act on the protection of the marine environment.

equivalent components. If, however, these options are not possible, the question arises if and how "equivalent restoration" of that kind of damage is possible and/or desirable in the context of the Act on the protection of the marine environment and (more generally) in the context of Belgian liability law. As already mentioned, this comes to the question as to whether non economically quantifiable losses in the marine environment (*e.g.* damaged ecosystems or destroyed wild fish grounds or sea fauna) can or have to be compensated for by a certain amount of money. This question has to be answered in the context of the Act on the protection of the marine environment on the one hand, and in the context of the common Belgian liability law on the other hand.

2.3.2. Valuation of ecological damage

Use values and passive values. Before answering the question as to whether a financial compensation for ecological damage would be possible and/or desirable in Belgian liability law, the question of the valuation of ecological damage has to be dealt with first.

In the literature concerning the different methods to value ecological damage, several kinds of values of natural goods are distinguished. Consequently, a natural good could have a use value and passive values (existence value and intrinsic value)[24]. The notion of use values indicates the value of natural goods by the use of them by people (*e.g.* the use of fish in a lake for fishermen or the use of a wood for hunters). Next to these use values, the notion of passive (or "non use" -) values indicates the value of the good by their existence. Even if some parts of nature are not used by people, they still have a certain value[25]. However, while the financial valuation of use values of natural goods is relatively simple (for it concerns commercial values of these goods, *e.g.* timber or commercially traded fish), it is very difficult to value passive values[26].

Although the notions of 'use values' and 'passive values' are not used in the Belgian law order, the non-economically quantifiable losses in the marine environment can be equated with the deterioration of the existence value of elements of the marine environment.

Valuation methods. Several theoretical methods are used to measure the passive values of natural elements[27]. The most important methods are the *hedonic property*

---

[24] CROSS, F.B., Natural Resource Damage Valuation, *Vanderbilt Law Review*, 42 (2), 280-297 (1989).

[25] CROSS, F.B. (1989), *l.c.*, 285-286; LARSSON, M.L. (1999), *The Law of Environmental Damage – Liability and Reparation*, Den Haag – London – Boston, Kluwer Law International and Stockholm, Norstedts Juridik, 551.

[26] For there is no method to value these values, which is based on real actions of people. See LARSSON, M.L. (1999), *o.c.*, 551 and CROSS, F.B. (1989), *l.c.*, 285.

[27] For an overview of the several valuation methods, see CARETTE, A. (1997), *o.c.*, 168–189; LARSSON, M.L. (1999), *o.c.*, 552–553.

prices, the *travel cost method* and the *contingent valuation method*. The hedonic property prices method tries to measure the implicit price of the 'environmental' characteristics of certain goods by examining the markets, where these goods are traded. For instance it is clear that the value of properties is partly dependent on environmental factors (degree of air pollution, noise pollution, etc.). The travel cost method tries to determine the value of nature reserves by examining the expenses of visitors of the reserves. The contingent valuation method creates an artificial market to value a natural good (that has no commercial value)[28]. People are asked which financial value they would give to certain environmental goods. Through a poll, people are asked their willingness to pay to realize a certain level of nature conservation on the one hand and, on the other hand, the amount of money they would like to get if a natural reserve would be lost (willingness to accept). This method is the only method which is able to measure the passive values of natural goods. However, the use of this method is controversial, due to its shortcomings, which can lead to unreliable results[29].

Besides these theoretical valuation methods, a few 'standard' valuation methods are used to value ecological damage. These methods use fixed rates to determine the compensation for damaged natural goods. In Alaska and the Russian Federation, a standard valuation system exists to determine the compensation for marine pollution[30]. In the Flemish Region, too, a standard valuation system is used to determine the compensation for damages to trees. This system intends to determine a correct valuation of trees, taking into account the several functions (including the aesthetic and ecological function), the species and the girth of the tree[31].

---

[28] See BATEMAN, I. , & WILLIS, K. (1999), *Valuing Environmental Preferences, Theory and Practice of the Contingent Valuation Method in the US, EU and Developping Countries*, Oxford, Oxford University Press, 645 p. See also *infra* in this book NUNES, P.A.L.D. & DE BLAEIJ, A.T., *Economic assessment of marine quality benefits: applying the use of non-market valuation methods* (Part III - Chapter 7) and VAN BIERVLIET, K., LE ROY D. & NUNES, P.A.L.D., *A contingent valuation study of an accidental oil spill along the Belgian coast* (Part III – Chapter 8).

[29] The method can lead to unreliable results due to polls, which are not executed in a correct way (unclear information). Furthermore, these studies are based on a fully hypothetical situation and the people questioned are not familiar with the financial valuation of natural goods. See CROSS, F.B. Restoring Restoration for Natural Resource Damages, *Toledo Law Review*, 319 (1993). One should also be vigilant for strategic answers, for the people questioned could possibly try to manipulate the results, knowing that it is a hypothetical situation and they will not really have to pay. See CROSS, F.B. (1993), *l.c.*, 316 and CARETTE, A. (1997), *o.c.*, 178.

[30] KOLODKIN, A., KISELEV, V. & KOROLEVA, N., Some new trends in legislation of the Russian Federation and its attitude towards conventions with regard to marine pollution in DE LA RUE, C. (1993), *Liability for damage to the marine environment*, London, Lloyds of London Press Ltd., 36. These systems determine the compensation by multiplying the amount of damaged goods (surface, quantity) by a fixed amount of money.

[31] It concerns the "Uniforme methode voor waardebepaling van de straat-, laan- en parkbomen behorend tot het openbaar domein" (http://www.vvog.org). For a review of this method, see BOCKEN, H., TRAEST, I. & DE JAGER, L. (1992), *Bomen in het recht, Een overzicht van het recht in het Vlaamse Gewest van toepassing op bomen*, Kluwer, 144–148.

Furthermore, a standard valuation method is used to value the compensation for the moral damage, caused by physical injuries caused by traffic accidents[32].

2.3.3 Monetary compensation for ecological damage in the context of Belgian liability law?

Common Belgian liability law. In common Belgian liability law the notions of 'use values' and 'passive values' are not used. It only distinguishes damage to appropriated goods and damage to non-appropriated goods. However, the non-economically quantifiable losses in the marine environment can be equated with the deterioration of the existence value of environmental goods. Belgian liability law does not compensate for the existence value of goods. The only theory which comes close to this is the theory of the compensation for the 'loss of opportunity'. According to this theory, one could get compensation for a loss of opportunity, if it concerns a real opportunity and if the loss is certain[33]. This theory is primarily used in the medical and judicial context (fault of doctors and lawyers). It is very doubtful if this theory could be used to claim compensation for the loss of opportunity to use the damaged environmental good in the future, for the opportunity would probably not be qualified as real.

The Act on the protection of the marine environment does not oblige, nor excludes financial compensation for ecological damage. Neither are the parliamentary works clear on this point. On the one hand, the restoration in kind is clearly preferred and the American and Italian theoretical models to financially assess ecological damage are rejected. On the other hand, the concept of 'impairment of nature' does not exclude that, in the future, a (financial) compensation for non-economically quantifiable damage could be claimed[34]. On this basis, one could argue that, if restoration in kind or the replacement of the damaged good by an equivalent good is not (completely) possible, the State could claim a financial compensation for the (remaining) ecological damage[35]. However, since the damaged natural goods do not have an economic value, the assessment of such a financial compensation is very difficult. In fact, the "ecological function" (the so-called 'passive values') has to be compensated. In that respect, the Act on the protection of the marine environment states that the King *may* lay down criteria and rules for determining an incident of environmental disruption and for fixing the relevant indemnity[36]. Furthermore, if

---

[32] "Indicatieve tabel van forfaitaire schadevergoedingen bij verkeersongevallen", *R.W.*, 246–249 (1998-99), made by the judicial authorities (*Nationaal Verbond van de Magistraten van Eerste Aanleg* and het *Koninklijk Verbond van Vrede- en Politierechters*).

[33] SIMOENS, D., Schade en schadeloosstelling, in: X (1999), *Beginselen van Belgisch Privaatrecht. Buitencontractuele aansprakelijkheid*, Antwerpen, Kluwer Rechtswetenschappen België, 54–56.

[34] Notice of Understanding, *Parl. St.* 1997-98, 1695/1, 35.

[35] CARETTE, A. (1999), *l.c.*, 372.

[36] Article 40 of the Act on the protection of the marine environment. Article 40 says that the King may lay down criteria and rules to determine the damage; it is not an obligation. This means that judges do not have to wait for a Royal Decree to determine that kind of damage and financially assess this kind of damage.

this question has to be answered positively, the problem of valuation has to be solved. More specifically, it is questionable if the use of theoretical economic valuation methods is acceptable and desirable in Belgian liability law.

2.3.4. Use of theoretical valuation methods in Belgian liability law?

Valuation of ecological damage in Belgian jurisprudence. Currently, no binding guidelines or valuation methods are used to value ecological damage. Belgian jurisprudence does not use the abovementioned theoretical valuation methods. In practice, Belgian judges commonly value ecological damage *ex aequo et bono*[37]. The Belgian Supreme Court (*Court of Cassation*) accepts this technique only if the judge clearly indicates why the valuation technique, proposed by the claimant, is not acceptable and if the damage cannot be valuated otherwise[38]. Apart from the *ex aequo et bono* valuation of ecological damage, judges often grant a compensation of a so-called one 'symbolic' euro. This 'technique', which primarily aims to compensate for moral damage, is often used to address claims introduced by environmental organisations. However, in the light of the principle of the 'integral compensation'[39], this technique should be denied. Damage has to be compensated, even if the valuation of the damage is difficult. Furthermore, ecological damage may not solely be qualified as moral damage, as this results in very low compensations[40]. Finally, we remember that in the Flemish Region, a standard valuation method is used to valuate damage to trees[41]. This formula is commonly used, but not binding. Judges can perfectly decline reasonably to use the formula and decide to value the damage *ex aequo et bono*.

Introduction of theoretical valuation methods into Belgian liability law? Given the fact that the economical valuation methods are not used in Belgian jurisprudence, it is questionable if these methods would be acceptable in the context of the principles of Belgian liability law. According to jurisprudence of the Court of Cassation, the judge autonomously determines the damage, the extent of the damage and the amount of the compensation. However, his decision has to be duly motivated and has to stay within the scope of the conclusion of the concerned parties[42]. Consequently, theoretically judges can use theoretical economic valuation methods, if they motivate their usefulness in the case concerned.

However, it is doubtful if these methods are consistent with Belgian liability law. At the moment, only two theoretical valuation methods are used in Belgian liability law: the '*Indicative table of standard compensation in case of traffic accidents*'[43]

[37] DELODDERE, S., unpublished note on ecological damage.

[38] SIMOENS, D., *o.c.*, 71.

[39] Meaning that the victim has to get full compensation for the damage which he suffered.

[40] DELODDERE, S., unpublished note on ecological damage; CARETTE, A. (1997), *o.c.*, 87.

[41] *Vide supra* section 2.3.3 and footnote 31.

[42] SIMOENS, D., *o.c.*, 67.

[43] *Vide supra* section 2.3.3 , footnote 31.

and the '*Uniform method to valuate damage to trees on the public domain*'[44]. The first method compensates moral damage, caused by physical injuries by traffic accidents. However, ecological damage should not be equated with moral damage, for, in case of ecological damage, one tries to compensate for the lost values of the environment itself, while, in case of moral damages, the pain and suffering of people is compensated.
The Flemish environmental law does not mention much concerning the valuation of ecological damage either. Only in Flemish nature conservation law is an application of the valuation of ecological 'damage' found. Article 90*bis* of the Wood Decree[45] foresees in a duty of compensation in case of deforestation. This compensation is given in kind or through a monetary compensation that has to be paid to a Flemish environmental fund. The compensation duty was further elaborated by a Decision of the Flemish Government[46]. The Decision distinguishes an equivalent compensating forestation (compensation in kind) on the one hand and a wood conservation contribution on the other hand. According to Article 5 of the Decision of the Flemish Government, the wood conservation contribution equals the surface of the equivalent compensating forestation (in m²), multiplied by 2 euro. The extent of the equivalent forestation equals the surface of deforestation, multiplied by a compensation factor, which is dependent on the kind of the trees. Consequently, here, a kind of standard valuation is used, based on the surface and the kind of the destroyed trees. No other valuation methods are used in Flemish environmental law.

Consistency with European, international and other national systems. Having examined the acceptability and the consistency of monetary compensation for ecological damage in Belgian liability law, one should also examine if the monetary compensation of ecological damage would be consistent with developments on the European and international level. More specifically, this has to be examined in the context of the European Directive on environmental liability with regard to the prevention and remedying of environmental damage[47] and the compensation practice of the International Oil Pollution Compensation Fund. Furthermore, one should look to the experiences of the US environmental liability law (CERCLA and OPA) and the Italian Act nr. 348 of 8 July 1986 in this respect.

(i) The *European Directive on environmental liability* clearly prefers natural recovery of the damaged natural resources or impaired services to baseline condition. If this primary remediation does not result in the full restoration of the damaged natural resources to baseline condition, complementary remediation can be

[44] See *supra* section 2.3.3, footnote 32.
[45] Wood Decree of 13 June 1990, *Belgian Official Journal* 28 September 1990.
[46] Decision of the Flemish Government of 16 February 2001 on the compensation of deforestation and the exemption of the interdiction of deforestation, *Belgian Official Journal*, 23 March 2001.
[47] Directive 2004/32 of 21 April 2004 of the European Parliament and the Council on environmental liability with regard to the prevention and remedying of environmental damage, *OJ* L 143, 30 April 2004, 56–75, hereinafter referred to as '*The Directive on environmental liability*'.

undertaken. The purpose of complementary remediation is to provide a similar level of natural resources and/or services, including, as appropriate, at an alternative site, as would have been provided if the damaged site had been returned to its baseline condition. Compensatory remediation shall be undertaken to compensate for the interim loss of natural resources and services pending recovery. This compensation consists of additional improvements to protected natural habitats and species or water either at the damaged site or at an alternative site. It does not consist of financial compensation to people. Monetary valuation techniques are used solely to determine the extent of the necessary complementary and compensatory remedial measures, if the use of resource-to-resource or service-to-service equivalence approaches is not possible. Consequently, under the scope of the European Directive on environmental liability, monetary valuation techniques are not used to determine a monetary compensation of ecological damage, but only to determine complementary or compensatory remediation measures. Thus, monetary compensation of ecological damage is not preferred.

(ii) *The compensation practice of the International Oil Pollution Compensation Fund.* Under the scope of the International Convention on Civil Liability for Oil Pollution Damage and the Fund Convention of 27 November 1992[48], ecological damage is not compensated. Only remediation measures are compensated, if they are reasonable, proportional and opportune.

(iii) *American (federal) liability law with regard to liability for ecological damage.* In the American law order, certain governmental agencies have the right to claim compensation for damage to natural resources, on the basis of the Comprehensive Environmental Response, Compensation and Liability Act[49] and the Oil Pollution Act. CERCLA contains a strict liability rule for damage, caused by dangerous substances. Under CERCLA, damage to or loss of natural resources (and the reasonable costs to valuate this damage and costs of research on the influence of the contamination on the health of people) can be compensated. The Department of Interior has elaborated damage valuation rules. The remediation of the damage is aimed at the recovery of the natural resource. In addition to these recovery costs, interim losses can be compensated[50]. These interim losses can be compensated financially. Several theoretical valuation methods are foreseen to value the compensation: the 'travel cost method', the 'hedonic property prices' and the 'contingent valuation method'. However, the preparatory works of the European Directive on

[48] International Convention of 27 November 1992 on the Establishment of an International Fund for Compensation for Oil Pollution Damage, as amended. See also *infra* in this book NICHOLS, J., *Scope of compensation for environmental damage under the 1992 Civil Liability Convention and the 1992 Fund Convention* (Part II – Chapter 4).

[49] Hereinafter referred to as 'CERCLA'. CERCLA came into force in 1980 and was fundamentally reviewed in 1986 by the *Superfund Amendments and Reauthorisation Act.* See BRANS, E.H.P. (2001), *o.c.*, 79 and CARETTE, A. (1997), *o.c.*, 328.

[50] Interim losses consist of the loss of services of the natural good during its recovery.

environmental liability mention that the Department of Interior has decided to evolve to less controversial valuation methods[51].
The Oil Pollution Act[52] that came into force in 1990 also contains a strict liability system for oil spills coming from ships. Under OPA, damage to natural resources and damage to the existence value of natural resources can be compensated. The National Oceanic and Atmospheric Administration has elaborated damage valuation rules. The damage assessment system is more aimed at determining the costs of the remediation measures than at the monetary valuation of the damage. Monetary valuation techniques are solely used to determine the compensatory measures to compensate for the interim losses. They are not used to determine a certain amount of money, that can be claimed as a compensation for damage to natural resources[53].

(iv) *The Italian Act nr. 348 of 8 July 1986 on the Establishment of a Ministry of Environment and the promulgation of rules with regard to environmental damage*[54].
Under the scope of this act, damage to appropriated goods as well as damage to non appropriated goods can be compensated. If possible, the damaged goods have to be recovered. If natural recovery is not possible, a monetary compensation can be granted. The judge has to valuate the damage using reasonable criteria[55]. Hereby, the gravity of the fault, the recovery costs and the advantages caused by the infraction have to be taken in account. However, the Act does not mention the use of theoretical valuation methods.

3. CONCLUSIONS

The Act on the protection of the marine environment has brought the notion of 'ecological damage' within the scope of the liability regime with regard to the marine environment. The Act clearly prefers restoration in kind or the replacement of the marine environment by equivalent components. However, it is questionable if 'equivalent restoration' of ecological damage, which cannot be restored in kind, is possible or desirable in the context of the dispositions of the Act on the protection of the marine environment and (more generally) in the context of Belgian liability law. In fact, this concerns the question of the possibility of a monetary compensation for non economically quantifiable losses in the marine environment. Although the notions of 'use values' and 'passive values' (existence value and intrinsic value) are not used in the Belgian law order, the non-economically quantifiable losses in the marine environment can be equalized with the deterioration of the existence value of environmental goods.

---

[51] Notice of Understanding of the European Directive on environmental liability, 15.
[52] Hereinafter referred to as '*OPA*'.
[53] See BRANS, E.H.P. (2001), *o.c.*, 139.
[54] For a review of this Act, see CARETTE, A. (1997), *o.c.*, 449–452; BRANS, E.H.P. (2001), *o.c.*, 306 and 331–335; GIAMPETRO, F., La spécificité du dommage écologique en droit italien, in: X. (1991), *Le dommage écologique en droit interne, communautaire et comparé*, Nice, *s.n.*, 109–118.
[55] CARETTE, A. (1997), *o.c.*, 452; BRANS, E.H.P. (2001), *o.c.*, 333.

First of all, the existence value of (environmental) goods is not compensated in Belgian liability law. Moreover, the Act on the protection of the marine environment does not prescribe the monetary compensation of ecological damage, which cannot be compensated in kind. However, the Act does not exclude this possibility either. More than a question of legal possibility, the question of the introduction of a monetary compensation of ecological damage, concerns a question of desirability and consistency with common liability law. As the existence value as a kind of damage does not exist already in common Belgian liability law at the moment and the preventive effect of the possibility of a monetary compensation for this kind of damage can be doubted and as there are still problems with respect to the valuation of this kind of damage, it is, in our view, doubtful if it is desirable to financially compensate ecological damage, which cannot be restored in kind, via liability law. Indeed, other techniques can be used to internalise these kind of costs (*e.g.* via administrative or criminal fines).

If, however, one should prefer monetary compensation of ecological damage, which cannot be restored in kind, via the liability system of the Act on the protection of the marine environment, the problem of the valuation of this damage still has to be solved. The existence value of environmental goods can only be valued using theoretical economic methods (*e.g.* the *Contingent Valuation Method*). At the moment, these kinds of methods are not used in the Belgian law order. Today, only the 'standard' damage assessment is applied (more specifically, in the context of damages to trees and in the context of a compensation of the moral damage, caused by the physical injuries caused by traffic accidents). However, from a strictly legal point of view, such valuation techniques can be used, provided that the usefulness of the techniques is duly motivated.

Consequently, although financial compensation of ecological damage to the marine environment and the use of theoretical economic models to value this kind of damage is possible under the scope of the Act of 20 January 1999 on the protection of the marine environment, it is questionable if it is desirable to compensate ecological damage, that can not be restored in kind, financially via liability law.

REFERENCES

- BATEMAN, I. & WILLIS, K. (1999), *Valuing Environmental Preferences, Theory and Practice of the Contingent Valuation Method in the US, EU, and Developing Countries*, Oxford, Oxford University Press, 645 p.

- BOCKEN, H., The compensation of ecological damage in Belgium, in: WETTERSTEIN, P. (1997), *Harm to the environment: the right to compensation and the assessment of damages*, Oxford, Clarendon Press, 143-158.

- BOCKEN, H. (1979), *Het aansprakelijkheidsrecht als sanctie tegen de verstoring van het leefmilieu*, Brussel, Bruylant, 500 p.

- BOCKEN, H., TRAEST, I., DE JAGER, L. (1992), *Bomen in het recht, Een overzicht van het recht in het Vlaamse Gewest van toepassing op bomen*, Antwerpen, Kluwer, 168 p.

- BRANS, E.H.P. (2001), *Liability for damage to public natural resources. Standing, damage and damage assessment*, The Hague/London/New York, Kluwer Law International, 457 p.

- CARETTE, A., De aansprakelijkheidsregeling uit de wet ter bescherming van het mariene milieu, *T.M.R.*, 362 – 374 (1999).

- CARETTE, A. (1997), *Herstel van en vergoeding voor aantasting aan niet-toegeëigende milieugoederen*, Antwerpen, Intersentia Rechtswetenschappen, 669 p.

- CROSS, F.B., Natural Resource Damage Valuation, *Vanderbilt Law Review*, 42 (2), 269-341 (1989).

- CROSS, F.B., Restoring Restoration for Natural Resource Damages, *Toledo Law Review*, 319 (1993).

- DELODDERE, S., not published note on environmental damage

- GIAMPETRO, F., La spécificité du dommage écologique en droit italien, in : X. (1991), *Le dommage écologique en droit interne, communautaire et comparé*, Nice, s.n., 93-106.

- HANOTIAU, M., La responsabilité en matière de troubles de voisinage, *Bull. Ass.*, 365-392 (1981).

- KOLODKIN, A., KISELEV, V. & KOROLEVA, N., Some new trends in legislation of the Russian Federation and its attitude towards conventions with regard to marine pollution in DE LA RUE, C. (1993), *Liability for damage to the marine environment*, London, Lloyd's of London Press Ltd., 268 p.

- LARSSON, M.L. (1999), *The Law of Environmental Damage - Liability and Reparation*, Den Haag – London – Boston, Kluwer Law International en Stockholm, Norstedts Juridik, 681 p.

- SCHRYVERS, J., Het nieuwe indicatieve tarief van de vergoedingen wegens lichamelijke schade, *R.W.*, 738–743 (1998 – 99).

- SIMOENS, D., Buitencontractuele aansprakelijkheid, Deel II, Schade en schadeloosstelling, in: X. (1999), *Beginselen van Belgisch Privaatrecht, XI*, Gent, Story-Scientia, 435.

- STIJNS, S. & VUYE, H., Burenhinder, in: X. (2000), *Beginselen van Belgisch privaatrecht. Zakenrecht. V*, Antwerpen, Story Scientia, 532 p.

- VAN OEVELEN (1984), *Schadevergoeding bij rechtmatige overheidsdaad in België*, Zwolle, 210 p.

- YANG, E.J., Valuing Natural Resource Damages: Economics for CERCLA lawyers, *Environmental Law Reporter* XIV (8), 10.311-10.317.

CHAPTER 3

COMPENSATION FOR ECOLOGICAL DAMAGE AND LATVIAN LAW

LAILA MEDIN, LMM

Deputy Head of European Affairs Office

State Chancellery, Republic of Latvia
Brivibas blvd 36, Riga, LV 1520 Latvia
E-mail: laila.medin@mk.fov.lv

1. INTRODUCTION

In this chapter the Latvian law regulating polluter's liability for maritime ecological damage is analyzed. The objective is to show how the Latvian legislator has regulated the liability of a polluter in case of marine environment pollution and to analyze how it has been applied in practice.

2. LEGAL FRAMEWORK

At the beginning it has to be stressed that the legal framework in this area of law has been influenced by two main distinctive sources of law:

i) general principles of environment protection designed for the needs of inland environment,

ii) international conventions of IMO[1] dealing with liability of ship and cargo owners for marine pollution.

---

[1] International Maritime Organization, specialized agency of United Nations see more http://www.imo.org/index.htm.

F. Maes (ed.), Marine Resource Damage Assessment, Liability and Compensation for Environmental Damage, 43–55.

As a result the rules applicable in case of an accident causing marine pollution are complex and they have created a legal regime, which has a lot of different aspects and which sometimes are not completely compatible with each other.

Based on these considerations this chapter has been divided into three parts in the following way:

i) an introduction with respect to each of the law and regulations applicable to an event of marine pollution;
ii) a description of how the different rules of these acts interrelate and apply to an theoretical accident;
iii) an examination of an actual application of these laws with respect to a concrete case, which took place in the port of Riga.

3. APPLICABLE LAWS AND REGULATIONS

In this section a short explanation of the content of the relevant laws and regulations in force in Latvia will be given. It is of particular importance because some of the rules are quite specific and they are not widely known.

The legal framework is built upon three laws and one Governmental regulation, *i.e.*:

1) the Environmental Protection Law[2], which is a framework law for environmental protection in Latvia;
2) the Maritime Code[3], which is a law covering, *inter alia*, marine environment protection, and which operates as a tool to implement international conventions stemming from the IMO;
3) the Law on Tax for Natural Resources[4], which is a financial instrument to fund environmental protection measures from the state budget; and
4) the Regulations of the Cabinet of Ministers on Calculation and Payment of Tax for Natural Resources[5], which is a tool to apply the Law on Tax for Natural Resources.

Soviet law-making, which might be of exotic interest, has more indirect than direct influence since all modern Latvian laws have been drafted after the breakdown of the Soviet era. So the influence is more felt due to the fact that the drafters can still be the same people who also did the same job during Soviet era.

---

[2] Likums par vides aizsardzību, approved by Supreme Council of the Republic of Latvia on 1991.08.06., Ziņotājs 33 1991.
[3] Jūras kodekss, approved by Saeima (Parliament) of the Republic of Latvia on 2003.05.29., Vēstnesis 91 2003.06.18.
[4] Likums par dabas resursu nodokli, approved by Saeima of the Republic of Latvia on 1995.09.14., Vēstnesis 152 1995.10.04.
[5] Dabas resursu nodokļa aprēķināšanas un maksāšanas kārtība, Regulations no.224 approved by the Cabinet of Ministers of the Republic of Latvia on 2002.06.18., Vēstnesis 97 2002.06.28.

3.1. Laws related to inland territory

The first and the third law mentioned above are structured and designed to fit the needs of the protection of the environment on the land territory of the country. They, of course, apply to pollution of the marine environment, but they do not treat marine pollution in any way different from pollution which can take place on the land. Therefore, when applying the rules of these laws to marine pollution a certain creative interpretation is necessary.

3.2. Maritime Code

The Maritime Code, by contrast, is heavily influenced by international conventions well known in the maritime community. The method of implementation of the conventions is based on the Scandinavian approach. The reason for using this method was to include the Latvian maritime legislation in the international circulation. The Ministry of Transport - after analyzing the different approaches present in the world - chose the Scandinavian Maritime Code system since their legal thinking is the closest to that of Latvia. More specifically the experts from the Ministry draw their inspiration from the Norwegian Maritime Code. But without doubt the Latvian Maritime Code cannot be the same as the Norwegian one, therefore, drafting of the Code demanded a lot of fine-tuning and adaptation efforts in order to achieve the result which:

1) on the one hand provides correct and internationally accepted ways of implementing international maritime conventions; and
2) on the other hand is compatible with the Latvian national legal system, especially with civil procedural rules.

3.3. Criminal and administrative law

There are two general laws, which are not specifically designed for environmental protection but which have a rather important role to play when analyzing the liability of the polluter. These are the Criminal Law[6] and the Administrative Violation Code[7]. The laws set out what actions are considered to be a criminal offence and in which cases the administrative fine can be imposed on a person linked with marine pollution.

4. ENVIRONMENTAL PROTECTION LAW

The Environmental Protection Law sets out the environmental policy guidelines, and confers powers to certain Governmental bodies to implement the policy. The law is general in its nature and the Governmental bodies working in accordance with this law consider it declarative and lacking a pragmatic approach. The norms of this law

---

[6] Krimināllikums, approved by Saeima of the Republic of Latvia on 1998.06.17., Vēstnesis 199/200 1998.07.08.

[7] Administratīvo pārkāpumu kodekss, approved by the Supreme Council on 1984.12.07, Ziņotājs 51 1984.

with respect to marine pollution have never been tested before the court because no serious accidents causing marine pollution have taken place in the territory of Latvia since the law came into force. The situations involving marine pollution have had a minor effect on nature and they have not caused noticeable damage or losses of the coastline or loss of fish stock, therefore no application of the law has occurred and only a theoretical analysis of the law is possible.

The main Articles of the Law are Article 3 (principles of environmental protection), Articles 7-10 (institutional framework), Articles 11-15 (the rights of citizens to live in qualitative environment), Articles 48-52 (liability of polluter) and Articles 53-56 (obligations of a polluter). The following sections include more detailed description of the relevant norms of the Law.

4.1. Article 3 of the Law

Article 3 of the Law contains principles of environmental protection, which should be observed throughout the application of environmental law. These principles are as follows:

- i) the principle of sustainable development (this principle serves as a guidance for Latvian state environmental policy),
- ii) the polluter pays principle (polluter pays for evaluation, limitation and prevention of pollution),
- iii) the principle of precaution (duty to chose methods less harmful to the environment when carrying out any kind of activities), and
- iv) the principle of evaluation before action (no activity can be approved without evaluating its effect on the environment and, as a result, the activity may not be implemented if damage to the environment is greater than the expected benefits from the activity).

Thereby the Article establishes the general principles of environmental liability setting that the person (either a natural person or a legal entity) who has caused the pollution has to compensate expenses and losses incurred due to pollution.

The Law does not specify the basis for establishing the link between the pollution and the polluter, that is, is it based on ownership of the object that has caused the pollution or is it based on its operation (use). This is of particular importance for the maritime sector where the ship owner and its operator are seldom the same person. As a result other laws related to marine pollution follow different routes. The Law on Tax for Natural Resources provides that the operator is liable for pollution, but the Maritime Code states that the registered owner of a vessel is liable for marine pollution as required by the Civil Liability Convention (CLC).

4.2. Articles 7-10 of the Law

Articles 7-10 create the institutional framework for environmental policy-making and its practical implementation. As far as the marine environment is concerned the responsible ministry is the Ministry of Environment (responsible for marine environment policy making) and its Marine Environment Board (responsible for the actual implementation and enforcement of the policy). In addition to these general Articles, Article 44 of the Law gives the authority to the inspectors of the Marine Environment Board to control the waters of Latvia and in case of pollution to investigate the reasons for the pollution, impose administrative penalties as provided by the Administrative Violation Code and to calculate the tax payable for illegal introduction of polluting materials into environment. Thus, Article 44 establishes the limits of power of responsible authority to enforce marine environment policy. The existing limited practice has not shown any need to change these powers.

4.3. Article 11 and 15 of the Law

Article 11 sets out the rights of citizens of Latvia to live in a qualitatively good environment. As a complement to this right Article 15 provides the right of citizens to receive compensation for loss or damage caused by pollution (law protects health, life, interests or property of person). As a balance to the right to quality of life the law imposes an obligation not to pollute. This obligation applies to all Latvian citizens and is also extended to all foreign persons who have their business located in the territory of Latvia including its territorial waters and exclusive economic zone.

This Article sets out the principle of territorial location of pollution, which means that the law applies only in situations when activities causing pollution have actually taken place in the territory of Latvia. This principle is relevant when analyzing the application of the Law on Tax for Natural Resources, which will be examined at a later stage.

It is also very important in relations with Lithuania. The pollution, which has occurred in Lithuanian territorial waters, will always move towards Latvian waters (due to streams), but the agreement between Latvia and Lithuania regarding cooperation in marine environment protection matters is not get concluded. Therefore, in case of pollution caused by Butinge oil terminal (located in Lithuanian territorial waters near the Latvian maritime boundary with Lithuania) victims will have legal and practical difficulties in obtaining compensation from the polluter.

This principle, however, is not applicable regarding the liability of the ship and cargo owner for marine pollution as established by the international conventions (CLC and FUND Conventions) and incorporated in Latvian legislation, in particular in the Maritime Code.

4.4. Article 48-52 of the Law

Article 48 gives a general overview regarding different types of liability the polluter has to face in case of an accident, *i.e.*:

i) criminal liability,
ii) administrative liability,
iii) obligation to compensate the expenses caused by the pollution.

A more detailed description of these types of obligation is included in the following sections.

4.5. Article 53 and 56 of the Law

Article 53 of the Law describes public obligations of the polluter. They are as follows:

i) to eliminate or minimize the consequences of the pollution,
ii) to compensate costs necessary to renew the environment or create the environmental values which are comparable with the ones lost.

The amount due from the polluter is calculated and the activities to be performed by the polluter are set by a commission established by order of the Director of the Marine Environmental Board. The costs payable by the polluter contain the following elements:

i) costs of operations to renew the environment,
ii) costs of operations to create comparable environmental values,
iii) losses caused by non-reparable damage.

The persons who have rights to claim damages are (Article 56 of the Law):

i) legal and natural persons who have suffered due to pollution with respect to their health, life, interests and property;
ii) governmental bodies to protect general interests of the state.

5. MARITIME CODE

Specific regulations with respect to marine environment protection are contained in the Maritime Code, which, *inter alia*, regulates the implementation of the Civil Liability Convention and the implementation of the Fund Convention, as well as procedural aspects related to the application of the CLC and Fund Conventions. The current status of the HNS Convention, the Bunker Oil Convention and Latvia's approach regarding the latest EU initiatives will also be briefly explained.

As mentioned before, the Latvian Maritime Code has been worked out on the basis of the Scandinavian approach as to how to implement international conventions related to maritime legal matters. The Code is a new legal instrument – it was adopted on 29 May 2003 and entered into force on 1 August 2003.

5.1. Implementation of CLC and FUND Conventions

Chapter 15-18 of the Code deals with the matters connected to ship owners' civil liability. It includes, inter alia, Articles implementing the 1992 CLC and FUND Conventions. Article 79 of the Code states that the owner of the ship is strictly liable for damage caused by oil as defined by the CLC convention. Other Articles of the relevant chapters provide detailed implementation of the relevant provisions of the CLC convention's Articles. Limits of liability are set in Article 82 and the maximum amount of liability is 89.77 millions SDR per ship per accident.

As regards the FUND Convention Article 89 of the Code refers to the application of this convention in Latvia without a detailed rewriting of the Convention. The exception is Article 90 dealing with the contributions to the Fund and Article 92, which provides that the Fund has the right to participate in Latvian court proceedings.

Article 96 extends the application of the CLC Convention's main rules to ships which are not covered by the Convention itself as well as to drilling platforms. Rules of the CLC Convention are also applied to oil products outside the scope of the Convention. The rules so extended are:

i) strict liability of the owner;
ii) exceptions from liability;
iii) right to limitation of liability.

Article 323 of the Code gives a formula for calculating the tonnage of a drilling platform.

The novelty for Latvian law are special provisions included in the Maritime Code related to the procedural aspects of establishing a limitation liability fund and its distribution. Although Latvia has been a member of the CLC and FUND Conventions for several years, no special procedure was designed for the specific needs of these Conventions. This, however, has just been a theoretical problem because no claims based on the CLC rules have ever been submitted to Latvian courts. The adoption of these rules, of course, should be seen as a very good and welcome progress, but one would hope that there will be no need to test their operation in reality.

According to the rules of the Maritime Code mentioned above, the limitation fund is established by order of the court, which at the same time also appoints an administrator of the fund. The administrator is responsible for finding the compromise solution among the claimants on the one hand and the ship owner on the other hand regarding the settlement of the claims, which is then approved by order of the court.

5.2. Implementation of the HNS and Bunker Oil Conventions

The Code does not deal with matters related to the HNS and Bunkers Conventions, the reason being that Latvia is not a party to these Conventions. However, both Conventions have been included in the working agenda of the Ministry of Transport.

5.3. Effect of EU legislation

A huge effect on Latvia's legal and administrative system has been caused by Latvia's application to become a member state of the European Union. This has tightened the requirements and deadlines for implementation of maritime safety rules, especially now after the *Erika* and *Prestige* accidents. As the result of the events of September, 11 the other priority of the EU is the implementation of the ISPS Code, which sets down the security requirements for the ports and the ships.

The most recent developments, which are of great interest for Latvia, are the new regulation on phasing out single hull tankers[8] and the new directive on environmental liability[9] in case of marine pollution. This directive has actually caused Latvia to change priorities with respect to the implementation of the HNS and the Bunker Oil Conventions, because they are included in the Annex of this Directive.

6. LAW ON TAX FOR NATURAL RESOURCES

The explanation regarding the legal framework in Latvia with respect to liability for marine pollution is concluded with the description of the Law on Tax for Natural Resources and Regulations of the Government on Calculation and Payment of Tax for Natural Resources. The regulations ensure the implementation of the abovementioned law.

6.1. Application of Law on Tax for Natural Resources to maritime accidents

The aim of the law is to limit uneconomical use of natural resources as well as to limit pollution. The main purpose for drafting and adopting this law was to create a system for charging tax for use of natural resources, as well as for introducing polluting substances into nature in accordance to licenses issued by a relevant governmental agency. Only very few sentences are devoted to illegal or - in accordance with the terminology used in this law – accidental pollution.

---

[8] See also *infra* in this book GONSAELES, G., *The impact of EC decision-making on the international regime for oil pollution damage* (Part II – Chapter 6 – 2.3.4.).

[9] See also *supra* in this book BRANS, E.H.P., *Estimating damages under the 2004 EC Directive on Environmental Liability* (Part I – Chapter 1) and PROOT, L. & BOCKEN, H., *Environmental damage and Belgian law* (Part I – Chapter 2 – 2.3.3.).

Such an approach means that if an accident takes place and as a result of such accident a ship has polluted the territory of Latvia the polluter will be liable to pay tax for natural resources, which is calculated in accordance with the methodology described in detail in the governmental regulations mentioned above. The tax collected is paid into a special state budget, which forms part of a national budget and which is used only for activities related to the protection of the environment.

6.2. Relevant definitions contained in Law on Tax for Natural Resources

The territory of Latvia is defined as land, territorial waters and continental shelf. The status of the exclusive economic zone is somewhat unclear. This definition means that the law will apply and tax will be payable only when pollution occurs in this territory. The Law on Tax for Natural Resources itself does not contain a direct indication as to the limited territorial application; that is specified in the Law on Environmental Protection.

Important too is Article 4 of the Tax law, which defines the taxpayer. The Law says that the taxpayer is any person (natural or legal) who introduces polluting substances into the territory of Latvia. From this Article two important conclusions arise:

i) tax is payable only if a polluting substance is actually discharged into the territory of Latvia, but not if the pollution has reached Latvia by force of nature;

ii) the taxpayer is a person who actually has caused pollution, which in case of maritime pollution has been interpreted to mean the operator of the ship at the time of the accident causing pollution. That means that the relevant authority, which is responsible for the tax collection, has to investigate who is the actual operator; it cannot simply send the invoice to the registered owner of the ship.

6.3. Calculation of tax

The Law on Tax for Natural Resources sets the basic principles for calculation of the tax. In case of water pollution, the Annex 4 to the Law shall be applied. The Annex sets down different amounts of money a polluter has to pay depending on how dangerous the pollutant is. In case of an oil spill, the oil is classified as a dangerous material and introduction of one ton of oil into the water if it is done in accordance with the issued license is charged 8,000 Lats, which is about 12,500 euro.

Article 19 of the Law on Tax for Natural Resources provides that in addition to the tax mentioned above a person who has caused a non-permitted pollution (that is pollution as a result of accident) has to pay a penalty. The Tax law delegates the power to the government to establish a methodology for the actual calculation of the tax and the penalty.

It should also be mentioned that the tax has to be paid only for pollution, which has actually persisted in nature. A polluting substance, which has been collected as a result of cleaning operations, is not taxed. The polluter, however, as mentioned before, is liable to pay for operations carried out to collect the pollution. Taking into account the severity of the penalty for accidental pollution, this is a very good way to encourage persons who have caused pollution to do their utmost to minimize the amount of pollution remaining in nature.

6.4. Government regulations

The Regulations of the Government on Calculation and Payment of Tax for Natural Resources provide detailed rules for implementing the Law on Tax for Natural Resources. However, only a few Articles are relevant for this topic. These Articles are as follows.

Article 14 of the Regulations requires inspectors of relevant Governmental bodies to control the use of natural resources and pollution of environment. The general (or normal) rule is that the taxpayer himself calculates the tax to be paid and the inspector checks the conformity of the tax paid with the requirements of the law. That applies in case of use of natural resources and also in case of licensed pollution. In case of an accident, an inspector of the Marine Environment Board makes the calculation and sends it to the Tax Authority for collection of the tax.

Article 53 describes the methodology for calculating the tax and the penalty in cases when the law is breached. For illegal pollution caused by an accident or illegal discharge of polluting substances the formula of calculating the tax to be paid is as follows: basic tariff + additional tariff, which is 3 times basic tariff + penalty, which is 2 times the total tax due to be paid. As a result the following calculation is applied: the amount of tax to be paid per one ton of polluting substance is multiplied by the amount of polluting substance left in nature. The result is multiplied by twelve, thus giving the final amount of money to be charged from the polluter.

The annexes to the Regulations contain a detailed description of polluting substances as well as documents specimen used in tax administration.

Annex 4 of the Regulation contains tariffs for water pollution. The Annex defines the degree of danger attributed to different substances. Oil is classified as a dangerous substance and, as was said before, the basic tariff of tax for dangerous pollutants is 8,000 Lats per one ton. That means that each ton of oil, which has been illegally spilled into the marine environment and which then has not been cleaned, is taxed with the amount 8,000 x 1 x 12 = 96,000 Lats, which is about 150,000 euro.

These rules have been applied twice so far. Both cases are still pending and one of them will be examined at the end of this chapter.

7. INTERRELATION BETWEEN LEGAL ACTS

7.1. Legal regimes

In this sub-chapter the interrelation between all legal acts mentioned above will be explained. This interrelation provides a full picture of all liability regimes possible under Latvian law. Briefly they are as follows:

i) the Maritime code – liability of ship and cargo owner to compensate for pollution damage;

ii) the Environmental Protection Law – administrative liability (reference to the Administrative Violation Code) and criminal liability (reference to Criminal Law);

iii) the Law on Tax for Natural Resources and Regulations of the Government on Calculation and Payment of Tax for Natural Resources – payment of tax for illegal pollution of natural resources (water).

This short list shows what kind of liability a polluter is exposed to in case of a theoretical oil spill case.

Here, it has to be stressed once again that since serious accidents causing real damage to the environment have not happened in Latvia, this summary is theoretical and it is far from clear how all elements of liability would be enforced and compensation paid to victims of an ecological accident in a real situation and how the laws would interrelate in reality.

7.2. Modeling of liability

Taking into account the factual situation mentioned above, a legal model has been worked out, which describes the potential liability of a person who has caused marine pollution in Latvian waters.

First, the registered ship owner would have to compensate claims submitted by private and public victims as defined by the 1992 CLC and FUND Conventions and implemented in the Latvian Maritime Code as well as based on the general principles of the Law on Environmental Protection.

Secondly, as referred to in the Law on Environmental protection there can be cases of criminal and administrative liability. Latvian Criminal Law sets out the following offences, which cause criminal liability:

i) for causing pollution twice during one year or has caused serious damage to the environment – a guilty person can be sentenced to 6 years in prison (maximum) or to pay a monetary penalty. But the law is not clear who would be liable – a possible person could be a captain of ship;

ii) for not eliminating pollution damage – a guilty person can be sentenced to 1 year in prison;

iii) for not reporting marine pollution to the responsible authorities if it is a duty to do so under a law – a guilty person can be sentenced to 2 years in prison.

The Latvian Administrative Violation Code sets the following administrative violations – polluting the environment, not keeping relevant documents in good order and not reporting the pollution case. The captain or other officer responsible for the matter will be fined for such violation.

Thirdly, the operator of a ship would be liable to pay tax for those polluting substances, which have not been recovered during clean-up operations, which is 150,000 euro per ton in case of dangerous substances such as oil.

Based on that, it should be concluded that the Latvian legislator seems to have been rather creative and has actually done quite a lot to enact laws, which would encourage ship owners, captains, other officers and ship operators to take due care and avoid polluting the marine environment, and - if that did happen - to take measures to prevent and/or minimize further damage and losses.

8. CASE STUDY

The Chapter is concluded with a brief case study describing the legal consequences of oil pollution from the tanker *Aquarius*, which took place on 20 December 2002 in the port of Riga. The study contains:

i) an explanation of the facts;
ii) a calculation of tax for natural resources;
iii) a description of possible claims to compensate damage to the environment;
iv) an analysis of administrative/criminal liability.

On 20 December 2002 a terminal operator in the port of Riga reported to the Marine Environment Board a case of oil spillage from a ship in the terminal. When the inspectors arrived at the place they found that at the left side of the ship there was an oil spill polluting approximately 200 m^2 of the water. The shipmaster denied any connection with the pollution and therefore inspectors carried out an inspection of the ship. They also took a sample of oil products on board the ship and from the water.

On the spot the inspectors detected certain short comings with respect to the ship's documentation for which the captain was fined in accordance with Article 61 of the Administrative Violation Code.

The inspectors also inspected the territory of the terminal and cargo supply pipelines, and it was evident that no pollution had originated from land. The oil spill remained at the source of the pollution due to the ice conditions in the port. Based on the evidence collected by the inspectors it was established that the total amount of oil spilled into the water was around 6,300 kg.

On 9 January, the Marine Environment Board received test results, which clearly showed that spilled oil and oil on board the ship were identical. The pollution was spilled into the water in breach of the law. Therefore these activities were defined as illegal pollution in accordance with Article 4 of the Law on Tax for Natural Resources.

Part of the oil was collected but most of it remained in the water, which meant that the operator of the ship became liable to pay the tax. In accordance with the Governmental regulation mentioned before for polluting substances (oil) illegally spilled into the water and not collected, the operator of the ship was charged a tax amounting 923,328 Lats, which is about 1,442,700 euro. This calculation was sent to the Tax Authority; the settlement of the case is still pending.

No claim arising from the CLC Convention or the Maritime Code was submitted to the court as a result of this accident, because no serious damage to the marine environment was caused.

The criminal investigation was carried out having regard the fact that the ship operator and/or the captain did not inform the authorities about the oil spill and did not start the necessary clean-up measures, but the cases was closed. Now the administrative case is pending before the court. The outcome of the case will show the actual possibilities to apply the current maritime environment liability regime in Latvia. Hopefully, it will serve as an encouragement to care more about marine environment protection.

9. CONCLUDING REMARKS

At the very end of this Chapter it is necessary to stress that the Latvian legislator has put into force all possible laws on marine environment liability as is done also elsewhere in the world, but it is a fight against the consequences, not against the source of the problem. Of course, it is beyond doubt, that it is vital to have appropriate rules for protecting the interests of the victims as well as rules for punishing "guilty" ship owners and/or operators. But it seems that the right medicine is to tighten the requirements, which are directly aimed at minimizing the possibility of pollution. In the author's opinion attention should be paid to the following aspects:

i) to control ship owners in order to ensure quality shipping;

ii) to have administrative and technical capacity to react quickly and adequately in a case of accident likely to cause spillage of dangerous and harmful substances;

iii) to have the capacity to organize effective clean up operations and restoration of the environment.

PART II.

INTERNATIONAL AND REGIONAL FUNDS FOR COMPENSATION FOR DAMAGE TO THE MARINE ENVIRONMENT

CHAPTER 4

SCOPE OF COMPENSATION FOR ENVIRONMENTAL DAMAGE UNDER THE 1992 CIVIL LIABILITY CONVENTION AND THE 1992 FUND CONVENTION

JOE NICHOLS

Deputy Director/Technical Adviser International Oil Pollution Compensation Funds

Portland House, Stag Place, London SW1E 5PN, United kingdom
Tel.: + 44 207 592 7100, Fax: + 44 207 592 7111
E-mail : Joe_Nichols@iopcfund.org

1. INTRODUCTION

The present international regime of compensation for pollution damage resulting from a spill of persistent oil from a sea-going vessel constructed or adapted to carry oil in bulk as cargo (normally a tanker) is based on two international Conventions adopted in 1992 under the auspices of the International Maritime Organization (IMO). These Conventions are the 1992 Civil Liability Convention and the 1992 Fund Convention. Prior to these Conventions becoming widely adopted compensation for pollution damage was based on two earlier Conventions, the 1969 Civil Liability Convention and the 1971 Fund Convention. The 1971 Fund Convention is no longer in force.

The Civil Liability Conventions govern the liability of shipowners for oil pollution damage. The Conventions lay down the principle of strict liability of shipowners

F. Maes (ed.), Marine Resource Damage Assessment, Liability and Compensation for Environmental Damage, 59–66.

through a system of compulsory liability insurance, which entitles victims of pollution damage to claim compensation directly from the shipowner's insurer. Shipowners are normally entitled to limit their liability to an amount that is linked to the size of the ship involved in an incident.

The Fund Conventions, which are supplementary to the 1969 and 1992 Civil Liability Conventions, establish a system for compensating victims when the compensation available under the applicable Civil Liability Convention is insufficient.

Each of the Fund Conventions established an intergovernmental organisation to administer the compensation regime it created, known as the International Oil Pollution Compensation Funds 1971 and 1992 (IOPC Funds) or the 1971 and 1992 Funds. The Organisations have a common Secretariat based in London.

At the time that the 1971 Fund Convention ceased to be in force, in May 2002, the maximum amount of compensation available per incident was 60 million Special Drawing Rights (SDR)[1] (US$87.8 million), including the amount paid by the shipowner or his insurer under the 1969 Civil Liability Convention. The maximum amount available under the 1992 Conventions is 203 million SDR (US$297 million).

The 1992 Fund has an Assembly composed of representatives of all Member States. The Assembly is the supreme body governing the 1992 Fund and it holds regular sessions once a year. The Assembly elects an Executive Committee composed of 15 Member States whose main function is to approve settlements of claims. From time to time the Assemblies of the two Funds have established Working Groups with the mandates *inter alia* to formulate criteria to be applied for the admissibility of such claims within the scope of the Conventions. The Funds' policy on claims for environmental damage has been reviewed a number of times by such Working Groups whose recommendations were subsequently adopted by the relevant Assembly.

The 1992 Conventions cover compensation for

i) Costs of clean-up, including preventive measures
ii) Property damage
iii) Consequential economic loss
iv) Pure economic loss
v) Costs of reinstatement of the environment and post-spill studies

This paper focuses on the development of the IOPC Funds' policy with regard to environmental damage over the past 25 years, either a result of decisions of the

---

1 The unit of account in the Conventions is the Special Drawing Right (SDR) as defined by the International Monetary Fund. In this paper the SDR has been converted into US dollars at the rate of exchange applicable on 10 December 2003, *i.e.* 1 SDR = US$ 1,46370.

Funds' governing bodies in response to specific claims or through the establishment of Working Groups with the mandate to formulate criteria to be applied for the admissibility of such claims within the scope of the Conventions.

2. ABSTRACT QUANTIFICATION OF ENVIRONMENTAL DAMAGE

2.1. The Antonio Gramski incident

The first incident involving the 1971 Fund, the grounding of the *Antonio Gramsci*[2] off Ventspils, USSR in 1979, gave rise to the question of admissibility of claims for compensation for damage to the marine environment. A claim of an abstract nature for ecological damage was made by the USSR against the shipowner. Although compensation could not be sought from the 1971 Fund, the USSR being Party only to the 1969 Civil Liability Convention at that time, the claim was of considerable interest to the Fund since it competed with the claims submitted by the Swedish Government for the amount payable by the shipowner. The amount claimed had been calculated on the basis of a mathematical formula laid down in USSR legislation. In the light of this claim, the 1971 Fund Assembly unanimously adopted in 1980 a Resolution stating that '*the assessment of compensation to be paid by the International Oil Pollution Compensation Fund is not to be made on the basis of an abstract quantification of damage calculated in accordance with theoretical models*'[3].

Following the adoption of this Resolution, a Working Group established in 1980 examined the question of whether and, if so, to what extent a claim for environmental damage was admissible under the 1969 Civil Liability and 1971 Fund Conventions. The Working Group took the view, which was subsequently endorsed by the Fund Assembly, that compensation could be granted only if a claimant had a legal right to claim under national law and had suffered '*quantifiable economic loss*'.

2.2. The Patmos incident

The *Patmos* incident, in the Straits of Messina, Italy in 1985 resulted in a claim by the Italian Government for unspecified environmental damage. No documentation was provided in support of the claim or the basis on which the amount claimed had been calculated. The 1971 Fund rejected this claim, which subsequently became the subject of legal proceedings[4].

The Italian Government maintained that the damage was a violation of the right to sovereignty of the territorial sea of the State of Italy. The Italian Court of First

---

[2] IOPC Fund Annual Report 1988, 62.
[3] Resolution No. 3 of the 1971 Fund, FUND/A/ES.1/13.
[4] FUND/EXC.40/8, para 2.

Instance rejected the claim stating that this right was not one of ownership and could not be violated by acts committed by private citizens. The Court also held that the State had not suffered any loss. The Italian Government appealed and maintained that the claim related to actual damage to the marine environment and to actual economic loss suffered by the tourism and fishing industries. In the Government's view the claim was therefore not in contravention of the interpretation of the definition of 'pollution damage' adopted by the Assembly in its 1980 Resolution referred to above.

The Court of Appeal overturned the Court of First Instance's judgement and held that the shipowner and the 1971 Fund were liable for the damage covered by the claim. The Court of Appeal appointed three experts who considered that the fishing activities had suffered some damage as a result of the fishermen having been unable to fish for a certain period. The experts stated, nevertheless, that their conclusions were only hypothetical since they had not been confirmed by factual evidence. The Court of Appeal granted the State of Italy compensation for damage to the marine environment. However, since the total amount of the admissible claims (including the Italian Government's claim) did not reach the limitation amount applicable to the *Patmos* under the 1969 Civil Liability Convention, the 1971 Fund was not called upon to make any compensation payments and was therefore not entitled to appeal to the Supreme Court of Cassation on this issue.

2.3. The Haven incident

Another incident in Italy, involving the *Haven* off Genoa in 1991[5], also gave rise to claims for environmental damage. The tanker caught fire and suffered a series of explosions, and it was estimated that over 10 000 tonnes of oil was spilled, a significant amount of which stranded on the shorelines of the Italian Riviera. Oil also impacted shorelines of the Principality of Monaco and the French Riviera.

The Italian Government submitted claims in respect of temporary damage to the environment that would recover naturally. Claims were also included in respect of irreparable damage to the environment, although it was left to the Court to quantify that damage. A number of regional and local authorities also submitted claims for environmental damage. The 1971 Fund opposed these claims on the grounds that they related to non-quantifiable elements of damage to the environment.

The Italian Court of First Instance decided that the 1969 Civil Liability Convention and the 1971 Fund Convention did not exclude environmental damage. The Court held that only the State of Italy (and not local authorities) was entitled to compensation for environmental damage. The Court took the view that the environmental damage could not be quantified according to a commercial or economic evaluation and assessed the damage as a proportion, approximately 1/3, of

[5] IOPC Funds Annual Report 1997, 46.

the cost of the clean-up operations, which in the Court's view represented the damage which was not repaired by these operations.

The 1971 Fund lodged opposition against the decision. The Court had based its decision on certain provisions in an Act of 1986 that had created the Italian Ministry of Environment. The Fund maintained that the liability for environmental damage laid down in those provisions was not applicable in relation to the Fund, because that liability was based on negligence, and the compensation, according to the provisions, had to be assessed on the basis of the degree of fault of the wrongdoer, the profit achieved by the wrongdoer and the cost in respect of the restoration of the environment. The Fund argued that according to Italian case law and legal doctrine, the compensation awarded under this Act had the nature of a sanction and that the damage assessed was therefore punitive. In the Fund's view, the criteria for assessment were inconsistent with the position of the Fund under the 1971 Fund Convention. The Italian Government requested that the amount awarded for environmental damage be increased to that set out in its original claim.

In March 1999[6] an agreement on a global solution of all outstanding issues relating to the *Haven* incident was concluded between the Italian State, the shipowner/insurer and the 1971 Fund. Under this agreement, the parties undertook to withdraw all legal actions in the Italian courts. The courts were therefore not called upon to make a final decision on the admissibility of the claims for environmental damage. The amount subsequently paid by the 1971 Fund in compensation did not relate to environmental damage.

3. NEW DEFINITION OF POLLUTION DAMAGE

The 1992 Conventions contained a new definition of pollution damage codifying the 1980 Resolution adopted by the 1971 Fund Assembly. The relevant wording in the new definition is *'that compensation for impairment of the environment other than loss of profit from such impairment shall be limited to costs of reasonable measures of reinstatement actually undertaken or to be undertaken'*. In view of this more precise definition it was hoped that the difficulties encountered by the 1971 Fund in the above-mentioned incidents would not arise under the 1992 Conventions. A Working Group established in 1993 included in its mandate the development of criteria governing the admissibility of claims for environmental reinstatement measures for adoption by the 1971 and 1992 Fund Assemblies. The Working Group[7] concluded that measures for reinstatement of the environment should fulfil the following criteria in order to be admissible for compensation:

---

[6] 71 FUND/A/ES.4/16, para 8.
[7] FUND/WGR.7/21.

i) the cost of the measures should be reasonable;
ii) the cost of the measures should not be disproportionate to the results achieved or the results which could reasonably be expected; and
iii) the measures should be appropriate and offer a reasonable prospect of success.

The Working Group stated that the test of reasonableness laid down in the 1992 Civil Liability Convention should be the same as that adopted in respect of preventive measures, *i.e.* that the measures should be reasonable from an objective point of view in the light of the information available when the specific measures were taken. The Working Group considered that it would normally be necessary to carry out an in-depth study before any measures of reinstatement were undertaken, and that the cost of such studies should qualify for compensation provided that they fulfilled the requirements generally applied by the Fund in this regard.

4. CLAIM FOR THE ECONOMIC CONSEQUENCES OF ENVIRONMENTAL DAMAGE

The *Nissos Amorgos* incident in Venezuela in 1997[8] gave rise to a claim related to the economic consequences of pollution damage to the marine environment. The tanker ran aground in the Maracaibo Channel in the Gulf of Venezuela resulting in a spill estimated at 3 600 tonnes of crude oil. Smaller releases of oil also occurred after the vessel reached Punta Cardon in the eastern part of the Gulf of Venezuela.

A claim was presented by six shrimp processing companies and 2 000 fishermen in respect of a reduction in catches of shrimp in Lake Maracaibo in 1998 alleging that this was due to the oil spill from the *Nissos Amorgos* the previous year. The 1971 Fund accepted that, despite the fact that fishing operations had not been interrupted, there had been a significant downturn in shrimp supplies to the processing plants. However, there had been no contemporaneous evidence linking the alleged loss to the contamination, although it was known that oil had passed through the shrimp spawning grounds. The Fund took the view that in the case of fishery losses arising some time after a pollution incident it would be unreasonable to expect such data to be available. However, laboratory experiments had demonstrated that low concentrations of oil could affect the reproduction and feeding of shellfish and the survival of shrimps. No other factors had been identified that could have led to the downturn in shrimp catches. After having examined the opinions of experts, the Fund decided that the oil from the *Nissos Amorgos* was most probably a significant contributory factor and that the claim was therefore admissible in principle[9]. In quantifying the losses attributable to the oil pollution account was taken of normal fluctuations in catches from year to year.

---

[8] IOPC Funds Annual Report 1997, 106.
[9] IOPC Funds Annual Report 2000, 75-76.

5. FURTHER CONSIDERATION OF THE ADMISSIBILITY OF CLAIMS FOR ENVIRONMENTAL DAMAGE

In April 2000 a Working Group established by the 1992 Fund Assembly to assess the adequacy of the international compensation system gave further consideration to the question of environmental damage.

The Working Group considered a proposal to introduce the concept of compensation for environmental damage as a violation of collective property whereby compensation would be available to a Member State on the basis of international rights under other Conventions to which it was a Party, the amount of compensation to be based on the conclusions of environmental impact studies conducted in accordance with procedures adopted by the 1992 Fund. The Working Group also examined a proposal to change the 1992 Fund's policy on environmental damage so that compensation would no longer be limited to cases where the claimant had suffered economic loss and to allow compensation to be calculated using theoretical models.

These proposals were not accepted since it was considered that they went beyond the present definition of 'pollution damage' in the 1992 Conventions. It was agreed that an examination should be made of what could be achieved within the present definition of 'pollution damage' as regards the admissibility of claims for reinstatement of the environment and for costs of environmental impact studies. There was also support for considering the issue of environmental damage in depth in the longer term.

There was considerable support in the Working Group for the encouragement of scientifically relevant studies that assisted in determining whether or not reinstatement measures were necessary and feasible, thereby minimising the possibility of claims resulting from unnecessary and ineffective measures. Such studies would be most appropriate after major oil spills where there was evidence of significant environmental damage, although if a study demonstrated no significant long-term effects or that no reinstatement measures were feasible, this should not exclude compensation for the costs of the study.

It was considered that in order for such studies to provide reliable and usable information it was important that they were carried out with scientific rigour and balance. This could best be achieved through a committee or other body established within the affected Member State to design and co-ordinate the programme. There would be benefit in the Fund becoming involved in the planning and in establishing the terms of reference of any study, since this could help in ensuring that it did not repeat work already carried out elsewhere.

As regards reinstatement measures, the Working Group focused on the development of additional specific criteria, recognising that most major oil spills do not cause permanent damage to the marine environment due to its great potential for natural recovery. The aim of any reasonable measures of reinstatement should be to bring the damaged site back to the same ecological condition that would have existed had the oil spill not occurred, or at least as close to it as possible (that is to re-establish a biological community in which the organisms characteristic of that community at the time of the incident are present and functioning normally). Measures taken at some distance from, but still within the general vicinity of, the damaged area might be acceptable, so long as it could be demonstrated that they would enhance the recovery of the damaged components of the environment.

In addition to satisfying the general criteria applied to the admissibility of all claims for compensation under the 1992 Fund Convention, the following specific admissibility criteria were also developed in respect of reinstatement measures[10]:

i) the measures should be likely to accelerate significantly the natural process of recovery;
ii) the measures should seek to prevent further damage as a result of the incident;
iii) the measures should, as far as possible, not result in degradation of other habitats or in adverse consequences for other natural or economic resources;
iv) the measures should be technically feasible;
v) the costs of the measures should not be out of proportion to the extent and duration of the damage and the benefits likely to be achieved.

The 1992 Fund Assembly largely endorsed the Working Group's conclusions[11]. A new edition of the 1992 Fund Claims Manual was published in November 2002 to reflect the 1992 Fund's position on post-spill studies and reinstatement measures.

6. CONCLUSIONS

Although the international compensation scheme has tended to focus on the compensation of victims of the economic consequences of oil spills, the experiences over the last 25 years have shown that the Member States have been willing and able to adapt the international compensation to the needs of society, in particular as regards the impact of pollution on the environment. Whilst there is probably little that can be done to extend the scope of compensation for environmental damage within the current legal framework, the existing Working Group may consider at a later stage whether the 1992 Conventions should be amended so as to widen their application in this important area.

---

[10] 92 FUND/A.7/4 Annex
[11] 92 FUND/A.7/29 para. 6.4 – 6.6.

CHAPTER 5

THE CANADIAN SSOP FUND AND ENVIRONMENTAL DAMAGE ASSESSMENT (EDA) IN CANADA

K.A. MAC INNIS, QC*

Administrator of the Ship-source oil Pollution Fund, Canada

90 Elgin Street, Eight Floor, Ottawa, Ontario, K1A 0N5, Canada
Tel.: + 1 613 990 58 07, Fax: + 1 613 990 54 23
E-mail: ssopf@rogers.com

1. INTRODUCTION

Canada has shown considerable foresight over the years in fashioning a unique well-functioning domestic compensation regime for ship-source oil pollution.

A few decades ago in most countries the legal options for seeking compensation for marine oil pollution damage and the recovery of costs and expenses for clean-up and monitoring were limited.
The catalyst for a made-in-Canada solution occurred in 1970 when the tanker *Arrow* grounded on Cerberus Rock in Chedabucto Bay, Nova Scotia. After the *Arrow* incident, major amendments were made to the *Canada Shipping Act* (CSA).

---

* The author, Kenneth A. MAC INNIS, QC, of Bedford, Nova Scotia, was appointed to the public office of Administrator of the Ship-source Oil Pollution Fund of Canada by the Governor General in Council in 1998, and again in 2004. The author's views in this paper do not necessarily reflect those of any other person, organization or government.

F. Maes (ed.), Marine Resource Damage Assessment, Liability and Compensation for Environmental Damage, 67–84.

2. A CANADIAN SOLUTION

The new oil spill legislation in Part XX of the *CSA*, which was passed by the House of Commons on 1 March 1971, and by the Senate on 30 March 1971, became part of Canadian Law on 30 June 1971[1]. Predating the entry into force of the international 1969 Civil Liability Convention by more than four years, and the international 1971 Fund Convention by more than seven years, the new Part XX was one of the first national comprehensive regimes for oil spill liability in the western world.

The principal elements of Part XX were:

(i) establishing the strict liability of shipowners to be responsible for costs and damages for a discharge of oil;

(ii) allowing the shipowner, in certain circumstances, to limit his liability to an amount linked to the ship's tonnage;

(iii) creating a new Fund, the Maritime Pollution Claims Fund (MPCF), to be available for claims in excess of the shipowner's limit of liability; and,

(iv) giving the Minister of Transport the power to move or dispose of any ship and its cargo discharging or likely to discharge oil.

This regime was in place between 1971 and 1989.

A levy of 15 cents per ton was imposed from 15 February 1972, until 1 September 1976, and during that period a total of approximately $35 million[2] was collected and credited to the MPCF from some 65 contributors. Payers into the MPCF included oil companies, power generating authorities, pulp and paper manufacturers, chemical plants and other heavy industries.

In 1989 Canada decided to increase its oil tanker spill cover by becoming a Contracting State in the international regime, while modifying and continuing its domestic regime.

3. THE CANADIAN COMPENSATION REGIME

3.1. The Canadian SSOP Fund

The Ship-source Oil Pollution Fund (SSOP Fund) came into force on 24 April 1989, by amendments to the *CSA*[3]. The SSOP Fund succeeded the MPCF. The accumulated amount of approximately $150 million in the MPCF was transferred to the SSOP Fund. Effective 8 August 2001, the SSOP Fund is governed by Part 6 of the *Marine Liability Act (MLA)*[4].

---

[1] RSC 1970, Ch. 27 (2nd Supp.).
[2] All figures are in Canadian currency.

The SSOP Fund is liable to pay claims for oil pollution damage and for costs and expenses incurred for clean-up measures etc., including measures taken in anticipation of a discharge of oil from the ship, at any place in Canada, or in Canadian waters including the exclusive economic zone of Canada, caused by the discharge of oil from a ship. It is Canada's national Fund.

In addition, there is the international compensation regime that mutualizes the risk of pollution (persistent oil) from sea-going tankers. Currently, Canada is a Contracting State to the international 1992 Civil Liability Convention, and the international 1992 Fund Convention.

The SSOP Fund is intended to pay claims regarding oil spills from ships of all classes. It is not limited to sea-going tankers and their oil cargo and their bunkers. It also covers compensation for bunker and other spills from non oil tankers.

The type of oil covered by the SSOP Fund is also greater than under the Civil Liability and Fund Conventions. It is not limited to persistent oil and includes petroleum, fuel oil, sludge, oil refuse and oil mixed with wastes.

The SSOP Fund is also available to provide additional compensation (a third layer) in the event that compensation under the 1992 Civil Liability Convention (CLC) and the 1992 Fund Convention, with respect to spills in Canada from oil tankers, is insufficient to meet all established claims for compensation[5].

During the fiscal year commencing 1 April 2004, the maximum liability of the SSOP Fund is approximately $143.6 million for all claims from one oil spill. This amount is indexed annually.

The SSOP Fund is a special account established in the accounts of Canada upon which interest is credited monthly by the Minister of Finance. As at 31 March 2004, the balance in the SSOP Fund was some $330.7 million.

During the fiscal year commencing 1 April 2004, the Minister of Transport has the statutory power to impose a levy of 43.06 cents per metric tonne of "contributing oil" imported into or shipped from a place in Canada in bulk as cargo on a ship. The levy is indexed annually to the consumer price index. Since 1976 the Minister has not had to use this continuing statutory power.

3.2. Classes of claims

The classes of claims for which the SSOP Fund may be liable include the following:

---

[3] RSC 1985, 3rd Supplement, Ch. 6.
[4] SC 2001, Ch. 6.
[5] See Appendix A to this paper.

(i) claims for oil pollution damage[6];
(ii) claims for costs and expenses of oil spill clean-up, preventive measures and monitoring[7];
(iii) claims for oil pollution damage and clean-up costs where the cause of the oil pollution damage is unknown and the Administrator of the SSOP Fund has been unable to establish that the occurrence that gave rise to the damage was not caused by a ship.

3.3. Loss of income

A widely defined class of persons in the Canadian fishing industry may claim for loss of income caused by an oil spill from a ship and not recoverable otherwise under the *MLA*. In section 88 of the *MLA* dealing with claims for loss of income, "claimant" means:

a) an individual who derives income from fishing, from the production, breeding, holding or rearing of fish, or from the culture or harvesting of marine plants;

b) the owner of a fishing vessel who derives income from the rental of fishing vessels to holders of commercial fishing licences issued in Canada;

c) an individual who derives income from the handling of fish on shore in Canada directly after they are landed from fishing vessels;

d) an individual who fishes or hunts for food or animal skins for their own consumption or use;

e) a person who rents or charters boats in Canada for sport fishing; or

f) a worker in a fish plant in Canada, excluding, except in the case of a family-type co-operative operation that has a total annual throughput of less than 1,400 metric tons or an annual average number of employees of fewer than 50, a person engaged exclusively in supervisory or managerial functions.

---

[6] Oil pollution damage, in relation to any ship, means loss or damage outside the ship caused by contamination resulting from the discharge of oil from the ship.

[7] Measures taken to prevent oil pollution damage from the ship, including measures taken in anticipation of a discharge of oil from the ship, are covered, to the extent that the measures taken and the costs and expenses are reasonable, and for any loss or damage caused by those measures.

It is noted that the International Oil Pollution Compensation Fund has adopted a policy of paying some claims for pure economic loss (in addition to the normally compensable consequential economic loss). Pure economic loss in this context usually takes the form of claims for loss of income suffered by persons whose property has not been contaminated by an oil spill. In assessing claims for pure economic loss the International Fund generally uses a test of reasonable proximity[8].

3.4. Strict liability

The *MLA* makes the shipowner strictly liable (without negligence) for oil pollution damage caused by his ship, and for costs and expenses for clean-up and preventive measures to the extent that both are reasonable[9].

"Shipowner", in relation to any ship other than a Convention ship, means the person who has for the time being, either by law or by contract, the rights of the owner of the ship with respect to its possession and use.

"Reasonable" applies to both clean-up and preventive measures. There is no definition of "reasonable" in the *MLA*. It is noted that ITOPF says that for the 1992 Civil Liability Convention and the International Oil Pollution Compensation Fund "reasonable" is generally interpreted to mean "*that the measures taken or equipment used in response to an incident were, on the basis of an expert technical appraisal at the time the decision was taken, likely to have been successful in minimizing or preventing pollution damage*". The fact that the response measures turned out to be ineffective or the decision was shown to be incorrect with the benefit of hindsight are not reasons in themselves for disallowing a claim for the costs involved. A claim may be rejected, however, if it was known that the measures would be ineffective but they were instigated simply because, for example, it was considered necessary 'to be seen to be doing something'. On this basis, measures taken for purely public relations reasons would not be considered reasonable."

The shipowner is not liable under the *MLA* if he establishes that the occurrence

a) resulted from an act of war, hostilities, civil war or insurrection or from a natural phenomenon of an exceptional, inevitable and irresistible character;

b) was wholly caused by an act or omission of a third party with intent to cause damage; or

c) was wholly caused by the negligence or other wrongful act of any government or other authority responsible for the maintenance of lights or other navigational aids, in the exercise of that function.

---

[8] For that Fund's guidelines for assessing loss of income, see its website and document 92FUND/A.9/20.

[9] Section 51(1) *MLA*.

The liability and compensation for pollution provisions of the *MLA* do not apply to:

(i) a drilling ship that is on location and engaged in the exploration or exploitation of the sea-bed or its subsoil in so far as a discharge of a pollutant emanates from those activities;

(ii) a floating storage unit or floating production, storage and offloading unit unless it is carrying oil as a cargo on a voyage to or from a port or terminal outside an offshore oil field.

3.5. A fund of last resort

As provided in the *MLA*, in the first instance, a claimant can take action against a shipowner. The Administrator of the SSOP Fund is a party by statute to any litigation in the Canadian courts commenced by a claimant against the shipowner, its guarantor, or the 1992 IOPC Fund. The extent of the SSOP Fund's liability is stipulated in section 84 *MLA*.

3.6. A fund of first resort

The SSOP Fund can also be a fund of first resort for claimants, including the Canadian government.

As provided in section 85 *MLA*, any person may file a claim with the Administrator of the SSOP Fund respecting oil pollution loss or damage or costs and expenses, with one exception. A Response Organization (RO), established under the *CSA*, has no direct claim against the SSOP Fund[10].

3.7. Security for claims

The Administrator is empowered to commence an action *in rem* against the ship (or against the proceeds of sale, if the ship has been sold) to obtain security to protect the SSOP Fund in the event that no other security is provided. The Administrator is entitled to obtain security either prior to or after receiving a claim, but the action can only be continued after the Administrator has paid claims and has become subrogated to the rights of the claimant. The Administrator may claim such security in an amount not less than the shipowner's limit of liability under the *MLA*.

3.8. The rule of law

The Administrator, as an independent authority, has a duty to investigate and assess claims filed against the SSOP Fund. For these purposes, he has powers to summon witnesses and obtain documents.

The Administrator may either make an offer of compensation or decline the claim. An unsatisfied claimant may appeal the Administrator's decision to the Federal Court of Canada within 60 days.

---

[10] Response Organizations are private bodies established for the purpose of responding to oil pollution threats. Under Canadian legislation, each ship who enters Canadian waters must have an agreement in place with an RO.

Where a claim is filed with the Administrator a claimant is not required to satisfy the Administrator that the occurrence was caused by a ship, but the Administrator shall dismiss a claim if satisfied on the evidence that the occurrence was not caused by a ship.

When the Administrator pays a claim, he is subrogated to the rights of the claimant.

The Administrator must take all reasonable measures to recover from the shipowner, the International Fund, or any other person liable, the payments to claimants from the SSOP Fund. Shipowners will normally repay the Administrator only to the extent of their legal liability.

Thus, particularly in the investigation, assessment and payment of claims, the Administrator must act in accordance with the laws governing the operation of the SSOP Fund. He must not act arbitrarily or in deference to external policies contrary to Canadian Law.

The Administrator is the Canadian official who directs payments of domestic claims and authorizes and monitors payments of all Canadian contributions to the International Fund from the SSOP Fund.

The Administrator is wholly accountable to Parliament for all payments out of the SSOP Fund.

The Administrator holds office during good behaviour for a fixed term and, on the same principle as for the judiciary, is subject to removal by the Governor in Council for cause.

Parliament has stipulated that the Administrator shall not hold any office or employment inconsistent with his duties as Administrator. He is a statutory authority that must be independent of the Crown, oil/shipping/insurance interests, *etc.*

3.9. A pragmatic approach

In addressing claims, the Administrator recalls that the current statutory claims regime, on the principle of strict liability, has as its "four cornerstones":

- (i) all costs and expenses must be reasonable;
- (ii) all clean-up measures taken must be reasonable measures;
- (iii) all costs and expenses must have been actually incurred;
- (iv) all claims must be investigated and assessed by an independent authority (the Administrator).

Experience shows that the investigation and assessment of claims is expedited when claimants provide convincing evidence and written explanations. This includes various justifications by the On-Scene Commander (OSC) and proof of payment, etc. Detailed logs and notes by the OSC and others are invaluable in facilitating the settlement and payment of claims. It is essential that the measures taken and the costs and expenses incurred are demonstrably reasonable. The claim should be presented in a timely manner.

3.10. Environmental damage assessment and restoration

Canada is a Contracting State to the 1992 Civil Liability and the 1992 Fund Convention.

The Third Intersessional Working Group of the 1992 IOPC Fund considered issues of environmental damages under the 1992 Conventions and Protocols. This resulted in changes to the 1992 IOPC Fund's Claims Manual[11].

Compensation for environmental damages is dealt with under the *MLA*, the 1992 Civil Liability Convention, the 1992 Fund Convention[12], and the US *OPA 90*.

The 1992 Civil Liability Convention and the 1992 Fund Convention, in their definitions provide that "pollution damage" means [in part]:

> *"(a) loss or damage caused outside the ship by contamination resulting from the escape or discharge of oil from the ship, wherever such escape or discharge may occur, provided that compensation for impairment of the environment other than loss of profit from such impairment shall be limited to costs of reasonable measures of reinstatement actually undertaken or to be undertaken... ".*

The *MLA* (the SSOP Fund's governing statute) defines "oil pollution damage" as:

> *"...in relation to any ship, means loss or damage outside the ship caused by contamination resulting from the discharge of oil from the ship."*

The *MLA* provides:

> *"the owner of a ship is liable for oil pollution damage from the ship."*

The *MLA* further provides:

> *"If oil pollution damage from a ship results in impairment to the environment, the owner of the ship is liable for the costs of reasonable measures of reinstatement actually undertaken or to be undertaken."*

---

[11] 92 FUND/A.7/29-18 October 2002, pages 3-4 and Annex 1.

[12] See also *supra* in this book NICHOLS, J., *Scope of compensation for environmental damage under the 1992 Civil Liability Convention and the 1992 Fund Convention* (Part II – Chapter 4).

In the US, *OPA 90* provides for payment of natural resource damage claims from the Oil Spill Liability Trust Fund. Only designated Trustees may submit natural resource damages. Under US regulations the trustee may consider a plan to restore and rehabilitate or acquire the equivalent of the damaged natural resource.

The technically justified reasonable cost for reinstatement/restoration measures, for which compensation is available under the 1992 Civil Liability Convention and the 1992 Fund Convention, might equate to primary restoration under the US Natural Resource Damage Assessment regulations (NRDA). However, the further measure of *OPA* NRDA is:

(i) the diminution in value of those natural resources pending restoration, *plus*
(ii) the reasonable cost of assessing those damages.

The 1992 Civil Liability Convention and the 1992 Fund Convention do not, by their definition of pollution damage, cover this latter sort of compensation provided by the NRDA regulations or other theoretically based assessments of environmental damage.

In Canada it has been mooted by some that the definition of oil pollution damage in the *MLA* appears sufficiently broad to allow the Administrator of the SSOP Fund to entertain claims for environmental damages for a loss not tied to some identifiable economic consequence. In response, others argue that in light of the particular provision respecting liability for the costs of reasonable measures of reinstatement, it is quite clear that such "non-use value claims" are not provided for under the SSOP Fund's governing statute.

So far, neither the Canadian courts nor the Administrator of the SSOP Fund have considered the meaning of pollution damage in the governing statute, in this context.

It is noted that the Supreme Court of Canada has touched on the notion of compensation for non-pecuniary loss in the matter of *British Columbia v. Canadian Forest Products Ltds*[13].

4. CURRENT DEVELOPMENTS – EXTERNAL

4.1. US NRDA

The US NRDA regulations provide a process to assess injuries to natural resources and design an appropriate restoration plan.

In February 2001, a Memorandum of Understanding (MOU) was signed between the International Group of P&I Clubs and the US National Oceanic and Atmospheric Administration (NOAA). The aim of the MOU is to promote expeditious and cost-effective restoration of injured natural resources and services resulting from ship-source spills in the US, as authorized by *OPA 90* and determined by the NRDA Regulations.

[13] 2004 SCC 38, file number 29266. The link following provides access to this decision: http://www.lexum.umontreal.ca/csc-scc/en/rec/html/2004scc038.wpd.html

The parties to this MOU will meet every six months to facilitate a regular exchange of technical information, such as ITOPF and NOAA technical papers and publications. In the event of a ship-source oil spill in the US, early contact will be established between the relevant P&I Clubs, ITOPF, NOAA, and its co-trustees. ITOPF's role is to provide technical information and analysis.

4.2. Canada's Environmental Damages Fund (EDF)[14]

Prior to 1995, any judgments obtained from a court or monies obtained from settlements reached between parties involving the Canadian government had to be paid into the Consolidated Revenue Fund by virtue of the *Financial Administration Act*[15]. Consequently, those monies could not be used to assist in environmental restoration projects. A new policy was developed to avoid this problem. The Treasury Board of Canada in 1995 authorized the creation of a special holding account (*Environmental Damages Fund*) for the purpose of allocating court awards and settlements, as well as voluntary payments and international funds compensation, towards environmental restoration projects.

The object of the Environmental Damages Fund (EDF) is to assist in the rehabilitation of injured or damaged environmental or natural resources and to ensure that proposed projects to help rehabilitate the environment are cost effective and technically feasible.

For instance, after the Crown successfully prosecutes a polluter under certain federal environmental legislation and a fine is imposed, or in a case where the federal government commences civil litigation against the polluter and either negotiates or obtains a judgment from a court in relation to restoration of environmental damages both with respect to past and future damage, the court, the Crown and the defence can recommend that the monies obtained be placed into the EDF. However, cleanup costs, actual response costs and legal costs are specifically excluded from the EDF.

This approach is seen to be effective. At the March 2001 sessions of the Third Intersessional Working Group of the 1992 IOPC Fund, ITOPF presented its views on compensation for environmental damages under the international 1992 Civility Liability and Fund Conventions[16]. In its paper[17] ITOPF refers to other approaches by the USA and developments in the European Commission. ITOPF comments on the EDF managed by Environment Canada:

---

[14] WRUCK, H., QC (2002), *Overview of Canadian Environmental Legislation and Compensation for Environmental Damage*, presented at the EDF national workshop, *Towards a national Environmental Damages Fund Action Plan*, hosted by Environment Canada, Gatineau, Qc, 11-13 December 2002. See also Harry J. Wruck QC, *The Federal Environmental Damages Fund*, 5 C.E.L.R. (3d) 120.

[15] R.S.C. 1985, c. F-11.

[16] See also *supra* in this book NICHOLS, J., *Scope of compensation for environmental damage under the 1992 Civil Liability Convention and the 1992 Fund Convention* (Part II – Chapter 4) and *infra* GONSAELES, G., *The impact of EC decision-making on the international regime for oil pollution damage* (Part II – Chapter 6).

[17] 92FUND/WGR.3/5/2.

> *"The Environmental Damages Fund serves as a special trust account to manage monies that are received as a result of court orders, awards, out-of-court settlements, voluntary payments and, so it is stated, compensation provided through international liability regimes. The Canadian Courts are apparently able to use various Federal laws to direct money to the Fund, including the Canadian Environmental Protection Act, Migratory Birds Convention Act, the Canada Wildlife Act, the Fisheries Act and the Canada Shipping Act. The Environmental Damages Fund is used to remediate damages to the environment, including assessment or research and development work required to support such restoration efforts. Whilst monies received may not always be used to restore the damaged area in respect of which they were received, it is a requirement that any projects have to be in the region/community where the incident occurred. This initiative is seen as both an effective economic disincentive for illegal activities and as a means of providing compensation for environmental damage."*

One of the problems that arose after 1995 and to some extent is still the case today, is that courts and even government counsel are not familiar with the EDF. As a consequence, not a great deal of money has been paid into the EDF.

In the Atlantic Region of Environment Canada alone, as at November 2004 in excess of $650,000 has been contributed to the EDF and $450,000 dispersed for worthwhile restoration projects. A major part of that contribution is composed of proceeds obtained through quasi-criminal charges filed under the *Canadian Environmental Protection Act* and sections 32, 35 and 36(3) of the *Fisheries Act*.

As government officials, prosecutors, judges and defence counsel become more aware of the EDF it may become more utilized.

For instance, on 25 February 2002, a Nova Scotia Provincial Court judge imposed the country's highest ever fine - $125,000 – for pollution of coastal waters that are a haven to thousands of seabirds. In this case, the Philippine – registered ship *Baltic Confidence* was charged for discharging at least 850 litres of oil-mixed bilge water in December 1999, about 158 kilometres southwest of Halifax. In pleading guilty to the quasi-criminal offense, lawyers for Prime Orient Maritime of Manila said the company agreed to a penalty of $80,000 and a contribution of $45,000 to Canada's Environmental Damages Fund. The *Baltic Confidence* incident was the first time that a shipping firm paid into the EDF.

Another successful aerial surveillance mission occurred in March 2002, when a fishery patrol aircraft spotted an oil slick about 120 kilometres southeast of Halifax. The slick was reported to be 40 kilometres long and 15 metres wide. The oil trailed directly astern of the foreign-registered bulk carrier *CSL Atlas*. Subsequently quasi-criminal charges were laid and, after an agreement was reached between defence lawyers and federal Justice Department officials, a Nova Scotia Provincial Court judge imposed a fine of

$125,000 on 25 November 2002. The fine includes a $50,000 assessment that will go to the EDF toward dealing with environmental damages caused by marine pollution.

The important point to recognize with respect to the administration of the EDF is that it establishes clear criteria and standards that apply both to applicants and decision-makers in relation to the use of the Fund monies in respect to the restoration and projects. There really are three important principles running through the process. First, the restoration projects must be cost effective. Second, they must be technically feasible. Third, they must be scientifically sound before Fund monies may be used in that manner. To a large degree these important principles have been borrowed from American jurisprudence such as in the *Puerto Rico* v. *SS Zoe Collocotroni*[18] case, where the court refused to grant damages for the restoration of the environment, unless the government had a realistic plan in place to restore the environment to its pre-spill state.

5. ENVIRONMENTAL DAMAGE ASSESSMENT (EDA) IN CANADA

Until recently, the enforcement of environmental laws and regulations in Canada was established through a system of fines for quasi-criminal offences under various applicable legislation. The traditional problem associated with this technique was the lack of accepted methods to match costs with the damage that had occurred. Judges used to only apply the deterrence criterion when sentencing companies for environmental offences[19]. Environment Canada is developing a new approach – Environmental Damage Assessment or EDA - to help in the quantifying of such costs. EDA allows practitioners to determine the costs of environmental damages while taking into account the economic and social impacts linked to them. This technique may serve as a substitute for lawsuits originated by groups of persons in order to hold companies liable for the wildlife damages caused by their actions[20].

In Canada, a variety of environment-related legislation provides opportunities for the application of EDA protocols. The *Canadian Environmental Protection Act*[21] is the most important law in which EDA can be applied, but the *Fisheries Act*[22], the *Migratory Birds Convention Act*[23], the *Canada Wildlife Act*[24], the *Canadian Shipping Act*[25] and the *Arctic Waters Pollution Act*[26] also provide opportunities for the use of EDA.

EDA is a managing tool developed by Environment Canada used to provide compensation for damages caused to the environment. However, it can also be used as a

[18] *Puerto Rico* v. *SS Zoe Collocotroni*, 456 F.Supp 1327 (D.P.R. 1978).
[19] See specifically : *R.* v. *United Keno Hill Mines Ltd.*, (1980) 10 C.E.L.R. 43 (Terr. Ct.).
[20] See specifically : *Western Canada Wilderness Committee v. British Columbia (Minister of Environment and Parks)*, [1988] B.C.J., No. 436 (B.-C. S.C.) (Quicklaw).
[21] S.C. 1999, c. 33.
[22] R.S.C. 1985, c. F-14.
[23] S.C. 1994, c. 22.
[24] R.S.C. 1985, c. W-9.
[25] S.C. 2001, c. 26.
[26] R.S.C. 1985, c. A-12.

deterrent to polluters because they now know they can be held accountable for the damages they cause. Companies are thus encouraged to take better care of negligent behavior that could be a potential source of pollution.

As noted, above, in Canada, some of the money collected by the use of EDA may be managed by Environment Canada in the special trust account called the Environmental Damages Fund (EDF). However, all of the fines collected for quasi-criminal environmental offences do not go into the EDF.

5.1. How does EDA work? - Main guidelines

The main objective of EDA is to restore affected ecosystems or specific natural resources to their normal pre-incident state. Before the actual assessment begins, it is necessary for experts to set up the necessary background of EDA within the pre-assessment stage that happens once the experts get on site for the first time. The first step in the pre-assessment process will be the identification of the resource that is endangered by the incident. This will be achieved by defining the nature of the resource's exposure and recording any direct observations of injuries to the resource. The second phase is restoration planning where a formal assessment of the injury is made and possible alternatives of remedies are considered. Finally, the pre-assessment stage is completed by the undertaking of the restoration plan as quickly as possible to try to limit the damages.

Many factors are important in the first evaluation of the damage on site. These are the main details taken into account in the pre-assessment stage and that will later be crucial in the financial evaluation of the damage, specifically:

(i) when and where the event occurred (characterize the environmental setting);
(ii) possible source of the contaminant and the extent of the contamination;
(iii) identify the product involved;
(iv) volume released to the environment, and its physical and chemical properties; and
(v) potential impacted components of the ecosystem (*i.e.* species)[27].

From this initial assessment, experts determine if a full EDA investigation is needed and if there is a need to impute the direct costs of restoration to the polluter. Therefore, the main guidelines of EDA reside directly in the criteria used to determine its utility in a given situation. From then on, the process will obey strict protocols that will help quantify the damage done to the environment.

---

[27] ORGANISATION FOR ECONOMIC CO-OPERATION AND DEVELOPMENT (2002), *Application of Environmental Damage Assessment and Resource Valuation Processes in Atlantic Canada*, 11.

5.2. The EDA protocols

The first concrete use of EDA in Canada occurred in the Atlantic regions after a series of fish kill events and small scale oil spills. The importance of following a specific protocol for assessing the damages became ever more important because of the reality of the situation. The protocols serve three main purposes. First of all, they ensure the accurate collection of data linking the hazardous release to the environmental damages. After this process, protocols are also crucial in protecting the data collected in the field from the damaging effects of time. Finally, they ensure cohesion between the multiple players that interact in the EDA process.

When quantifying the cost of damages that will later be charged to a polluter, two approaches can be used by the EDA team depending on the given situation: Clement's proposal on EDA applied to small scale petroleum spills[28] and the Sawyer, Hundert and Macdonald approach to hazardous release of chemical agents[29].

5.2.1. Clement's proposal

Clement proposes a simplified approach to the assessment of costs due to environmental injuries when small scale petroleum spills take place. His technique limits the costs of extensive field investigations and focus on a low-cost approach to EDA. Usually, a complete EDA requires precise data collection and in-depth socio-economic analysis in order to minimize to the possibilities of errors. However, Clement's protocol provides a simplified evaluation and uses only three components in EDA: ecological costs, service flow costs and assessment costs. Clement believes in the reasonable trade-off provided by this simplified method because his protocol comes at the expense of some precision in evaluating damages but doesn't require costly and time-consuming research.

To clearly demonstrate the extent of his method, Clement applied this protocol to a small scale gasoline spill (less than 10 000 l) that affected a Nova Scotia cove in 2001. The area affected was 670 m^2 of shoreline which was comprised of man-made structures, cobble beach and mixed-sediment beach. Firstly, Clement used the benefits transfer economic valuation technique to evaluate the restoration costs of the spill. This implies the process of estimating the economic benefits of a resource in one location and using those estimates to value a resource in another location. Clement thus estimated that the ecological restoration costs of this particular spill were nearly $29,000. The service flow costs in this case were correspondent to recreational loses because of the lack of industries or businesses in the area that could have been affected by the spill. Using a national survey that measured the importance of nature to Canadians, Clement evaluated those costs at $1,300 for the period the coastline was oiled. Finally, the assessment costs were kept down to a reasonable $4,750 by Clement's protocol. In total, the small scale gasoline spill in the cove was estimated to cost nearly $35,000.

---

[28] CLEMENT, Ph.G. (2001), A Simplified Approach for Valuing Environmental Injuries from Small Petroleum Spills in Coastal Environments, in: *Proceedings from the Arctic and Marine Oil Spill Program Conference, hosted by Environment Canada* (Edmonton, June 2001).

[29] SAWYER, D., HUNDERT, E. & MAC DONALD, K. (2001), *Damage Assessment and Restoration Under Canadian Environmental Legislation: Applied Ecological Economics,* Presented at the *Annual Canadian Society for Ecological Economics (CANSEE) Conference*, hosted by McGill University, Montreal, Canada (24-26 August 2001).

5.2.2. The Sawyer, Hundert and Macdonald approach

This technique is mainly used in situations where wildlife has been affected by the environmental incident. In fact, this protocol was applied in the beginning stages of EDA in 1999 when a series of fish kill events occurred on Prince Edward Island. Its main difference with the Clement method resides in its first step: the evaluation of ecological costs. This approach takes the benefits transfer economic valuation technique and applies it to invertebrate organisms in the stream in order to get a precise assessment of the overall wildlife damage caused by the spill because these micro-organisms act as indicators of the quality of life of the body of water. By only considering the damages caused to those organisms, the experts can get a very good idea of the actual damage done to the whole ecosystem while reducing the assessment costs of the process.

5.3. Case Law and EDA

As mentioned above, Clement's simplified EDA approach was applied to a case that had already been through the Canadian justice system. The issue had already gone to trial and the judge in this case fined the responsible party $10,000, which is less than a third of what a conservative EDA process evaluated. The fine in this matter was determined according to case law and precedent, although most of the money went to the Environmental Damages Fund.

Two recent decisions were located where the court ordered the responsible party to contribute to the Environmental Damages Fund as part of the sentencing[30]. However, in both cases, the judge did not apply the EDA process nor did he rely on EDA expertise to determine the fine applicable to the respondent. The fines in these decisions were also evaluated according to the traditional use of case law and precedent.

5.4. An EDA case study

In light of the lack of case law that applies the EDA process, a report contributed by Environment Canada, as a national case study, to the organization for Economic Co-operation and Development (OECD) combined both Clement's protocol and the Sawyer, Hundert and Macdonald approach, in a recent essay on economic aspects of biodiversity. For this exercise, the authors used the contamination of the Valleyfield River in Prince Edward Island by pesticides. This incident killed at least twenty five hundred brook trout and a recreational fishery was closed as a result of this spill.

As Clement's protocol suggests, a three part evaluation of economic damages was made: one for the restoration cost of the trout population, one for the service costs induced by the incident, and one for the cost of the assessment process.

[30] *R.* v. *Canada (Ministry of Indian Affairs)*, [2000] O.J., No. 5076 (Ont. sup. Ct.) (Quicklaw) ; *R.* v. *White Pass and Yukon Corp.*, [1997] B.C.J., No. 3192 (B.C. prov. Ct.) (Quicklaw).

5.4.1. Ecological restoration costs

The first aspect of restoration in this case is the restocking of the river to compensate for the injuries to the ecosystem. In order to evaluate the cost of this process, an estimate of fish loss was used and market prices for hatchery trout were examined. Restoration costs in the first year would thus add up to $1,800. However, it is very likely that only a fraction of the dead fish was included in the analysis which could indicate that the figure is a conservative evaluation and could be much greater.

In addition to the restocking process, this ecosystem restoration program will involve the establishment of a fish stock monitoring program that will ensure the rebuilding of the population structure of the species. This activity will most likely require additional fish releases and would call for an investment of approximately $7,800, which brings the total cost of ecological restoration to $9,600.

5.4.2. Service flow costs

The spill in the Valleyfield river was the sole cause for the closing of the recreational fishery in the area. Therefore, in order to calculate the overall financial damage suffered by the fishery, consumer estimates from the preceding years were used to determine the number of participants that were likely to attend during that period. When combining this number to the total number of days the fishery may be closed due to the incident, it appears that the community would experience a total loss of between $10,700 and $14,200.

5.4.3. Assessment costs

The total cost of the environmental incident included the cost of people's professional time in the EDA process. Added to expenses were all the out-of-pocket costs such as travel, laboratory analysis, investigation materials, and disposal of dead fish. The total of these costs reached $15,300.

5.4.4. Total estimated costs

The EDA process in this case would have determined a total cost of at least $35,600 to $39,000 in economic damages. However, due to some of the approximations that have been made throughout the process, it must be understood that this figure would act as a starting point for assessment of damages associated with the fish kill.

6. CONCLUSIONS

Due to the infancy of the Environmental Damage Assessment process in Canada, it is clearly at a stage in its history where conflict emerges between the theoretical aspects developed by its creators and its use by judges. We have not been able to find cases where the process was taken into account by Canadian courts nor have we found specific legislation that addresses the issue of damage assessment. However, a number of authors recognize the use of EDA and highly recommend its application in the legal system, partly because of the gaps between fines and actual damage in environmental offences. The development of the Environmental Damages Fund by Environment Canada may prove to be a strong influence on judges to call upon EDA. The impact, if any, of such developments on the statutory civil Liability of the SSOP Fund remains to be seen.

REFERENCES

- CANADA, ENVIRONMENT CANADA (2001), *Framework for Environmental Damage Assessment and Restoration.*

- CLEMENT, Ph.G. (2001), A Simplified Approach for Valuing Environmental Injuries from Small Petroleum Spills in Coastal Environments, in: *Proceedings from the Arctic and Marine Oil Spill Program Conference, hosted by Environment Canada (Edmonton, June 2001).*

- FAIETA, M.D., KHON, H.B, KLIGMAN, R. & SWAIGEN, J. (1996), *Environmental Harm, Civil Actions and Compensation*, Toronto, Butterworths.

- HALLEY, P. (2001), *Droit pénal de l'environnement: l'interdiction de polluer*, Cowansville, Les Éditions Yvon Blais Inc.

- MC CONNEL, M.L. & GERLOCK, E.C. (1999), *Environmental Spills, Emergency Reporting, Clean-up and Liability*, Toronto, Carswell.

- ORGANIZATION FOR ECONOMIC CO-OPERATION AND DEVELOPMENT (2002), *Application of Environmental Damage Assessment and Resource Valuation Processes in Atlantic Canada*

- SAWYER, D., HUNDERT, E. & MAC DONALD, K. (2001), *Damage Assessment and Restoration Under Canadian Environmental Legislation: Applied Ecological Economics,* Presented at the *Annual Canadian Society for Ecological Economics (CANSEE) Conference*, hosted by McGill University, Montreal, Canada (24-26 August 2001).

- THOMPSON, G., MAC CONNEL, M.L. & HUESTIS, L. (1993), *Environmental Law and Business in Canada*, Ottawa, Canada Law Book Inc.

- WRUCK, H. (2002), *Overview of Canadian Environmental Legislation and Compensation for Environmental Damage*, presented at the EDF national workshop, *Towards a national Environmental Damages Fund Action Plan*, hosted by Environment Canada, Gatineau, Qc (11-13 December 2002).

- WRUCK, H. J., QC, *The Federal Environmental Damages Fund*, 5 C.E.L.R (3d) 120.

Appendix A

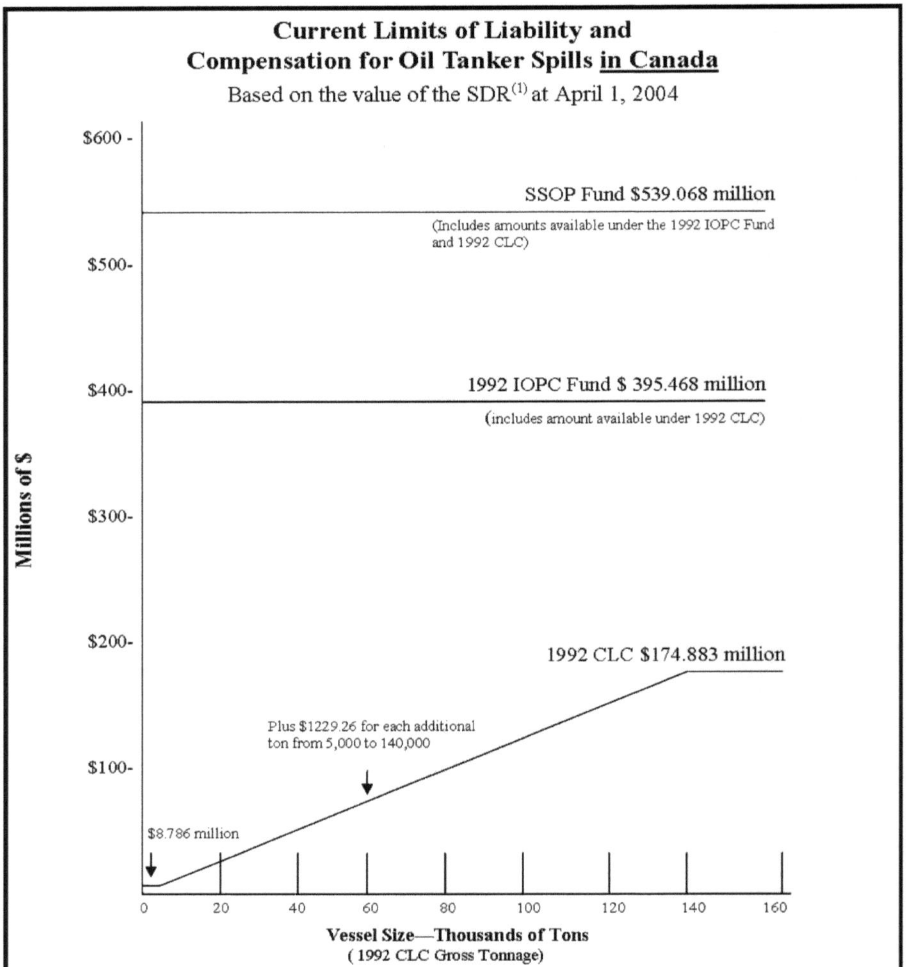

(1) The value of the SDR at April 1, 2004, was approximately $1.94812. This actual value is reflected above.

Appendix A shows the current limits of liability and compensation available under the 1992 CLC, the 1992 IOPC Fund Convention, and the SSOP Fund for oil spills from oil tankers in Canada, including the territorial sea and the exclusive economic zone. Because of the SSOP Fund, Canada has an extra cover over and above that available under the international Conventions.

N.B. The above aggregate amount available under the 1992 CLC and the 1992 IOPC Fund is $395.468 million effective November 1, 2003. The SSOP Fund amount of some $143.60 million on top of that, results in $539.07 million being available now for a tanker spill in Canada - without reference to proposed IOPC "optional" Supplementary Fund.

CHAPTER 6

THE IMPACT OF EC DECISION-MAKING ON THE INTERNATIONAL REGIME FOR OIL POLLUTION DAMAGE: THE SUPPLEMENTARY FUND EXAMPLE

GWENDOLINE GONSAELES, LL.M

Assistant, Public International Law Dpt. - Ghent University
Sr. Researcher, Maritime Institute - Ghent University

Universiteitstraat 6, 9000 Ghent, Belgium
Tel.: + 32 9 264 68 94 – Fax: + 32 9 264 69 89
E-mail: Gwendoline.Gonsaeles@UGent.be

1. INTRODUCTION

The legal provisions governing liability and compensation for oil pollution damage largely depend on international regulations, as laid down in international agreements that have been developed under the auspices of the International Maritime Organization (IMO). Nevertheless, the international liability and compensation regime for oil pollution damage is not to be seen as a static instrument: a wide variety of factors have influenced its establishment and development.

1.1. Leitmotiv: oil pollution incidents

International law - like all law - is to be developed to provide for an answer for recent evolutions and demands of society. Unfortunately, the need for a legal framework often occurs in the aftermath of regrettable events, generating pressure on the political willingness to create such a framework. The occurrence of some

F. Maes (ed.), Marine Resource Damage Assessment, Liability and Compensation for Environmental Damage, 85–131.

major oil spills and their negative consequences on the economic, social and ecological life in the region where the incident has taken place, has been the leitmotiv for the establishment and development of the international liability and compensation regime for oil pollution damage. Both the 1969 Civil Liability Convention and the 1971 Fund Convention[1] have been adopted as a direct consequence of the incident with the oil tanker *Torrey Canyon* (18 March 1967). The Conventions aim at ensuring the availability of adequate compensation to persons who suffer oil pollution damage resulting from maritime casualties involving oil-carrying ships[2]. This incident highlighted the risks of oil transport by sea, especially by means of *Very Large Crude Carriers* (VLCC) or by substandard ships, and emphasized the need for international regulations.
Subsequent oil pollution incidents, such as the incident with the *Amoco Cadiz* (16 March 1978), have clearly shown the need for higher compensation limits and raised questions as to the admissibility of claims for compensation. A first attempt to provide for higher compensation limits and to codify the practice of the International Oil Pollution Compensation Fund, 1971 (1971 IOPC Fund) towards compensation has been made with the adoption of the 1984 Protocols[3]. However, in the aftermath of the incident with the *Exxon Valdez* (24 March 1989), the United States decided to adopt national legislation - the *Oil Pollution Act 1990* (OPA '90) - and not to ratify the 1984 Protocols. As a consequence, the 1984 Protocols could never come into force, urging the international community to provide for the 1992 Protocols[4], which changed the entry into force requirements of the 1984 Protocols.

---

[1] The International Convention on Civil Liability for Oil Pollution Damage, 1969 (Brussels, 29 November 1969, entry into force 19 June 1975, *Ratification of Maritime Conventions*, I.7.30), puts a strict liability on the shipowner, up to a certain limit. The International Convention on the Establishment of an International Fund for Compensation for Oil Pollution Damage, 1971 (Brussels, 18 December 1971, entry into force 16 October 1978, *Ratification of Maritime Conventions*, I.7.90), provides for additional compensation in cases where the compensation available under the 1969 Civil Liability Convention was inadequate or unobtainable, to be paid by an international compensation fund, financed by the oil receivers. Both Conventions have been amended by the 1976 Protocols, changing the unit of account in which the liability and compensaton limits were expressed from poincaré franc to Special Drawing Rights (1976 Protocol to the International Convention on Civil Liability for Oil Pollution Damage, 1969 (London, 9 November 1976, entry into force 8 April 1981, *Ratification of Maritime Conventions*, I.7.40); and Protocol of 1976 to amend the International Convention on the Establishment of an International Fund for Compensation for Oil Pollution Damage, 1971 (London, 19 November 1976, entry into force 22 November 1994, *Ratification of Maritime Conventions*, I.7.100).

[2] The incident with the *Torrey Canyon* also raised legal questions as to the right of intervention of coastal states for oil pollution at sea, resulting in the adoption of the International Convention Relating to Intervention on the High Seas in Cases of Oil Pollution Casualties, 1969 (Brussels, 29 November 1969, entry into force 6 May 1975, *Ratification of Maritime Conventions*, II.7.70).

[3] 1984 Protocol to the International Convention on Civil Liability for Oil Pollution Damage, 1969 (London, 25 May 1984, *Ratification of Maritime Conventions*, I.7.50) and 1984 Protocol to amend the International Convention on the Establishment of an International Fund for Compensation for Oil Pollution Damage, 1971 (London, 25 May 1984, *Ratification of Maritime Conventions*, I.7.110).

[4] 1992 Protocol to amend the International Convention on Civil Liability for Oil Pollution Damage, 1969 (London, 27 November 1992, entry into force 30 May 1996, *Ratification of Maritime Conventions*, I.7.51) and 1992 Protocol to amend the International Convention on the Establishment of an International Fund for Compensation for Oil Pollution Damage, 1971 (London, 27 November 1992, entry into force 30 May 1996, *Ratification of Maritime Conventions*, I.7.111). The 2000 Protocol to the International Convention on the Establishment of an International Fund for Compensation for Oil Pollution Damage,

The 1992 protocols to the 1969 Civil Liability and 1971 Fund Conventions[5] widened the scope of the Conventions to cover pollution damage caused in the exclusive economic zone (EEZ) or equivalent area of a State Party. They also extended the Conventions to cover spills from sea-going vessels constructed or adapted to carry oil in bulk as cargo so that they apply to both laden and unladen tankers, including spills of bunker oil from such ships. The 1992 Protocols also codified the practice of the IOPC Fund 1971 towards compensation, especially regarding environmental damage. The compensation limits were those originally increased in 1984. The 1992 Protocol to the 1971 Fund Convention established a separate International Oil Pollution Compensation Fund, 1992 (1992 IOPC Fund). Some subsequent oil pollution incidents — namely with the *Aegean Sea* (3 December 1992), with the *Braer* (5 January 1993), and with the *Sea Empress* (15 February 1996) — stimulated a rapid entry into force of the 1992 Protocols (*i.e.* on 30 May 1996). Following the incidents with the *Nakhodka* (2 January 1997) and the *Erika* (12 December 1999), the IMO quickly reacted to calls for additional compensation by invoking the tacit acceptance procedure to raise the compensation limits as laid down in the 1992 Civil Liability and Fund Conventions[6].

1.2. The influence of domestic legislation

The original 1969 Civil Liability and 1971 Fund Conventions qualified "*pollution damage*" as "*(i) loss or damage outside the ship by contamination resulting from the escape or discharge of oil from the ship, wherever such escape may occur, and (ii) the costs of preventive measures and further loss or damage caused by preventive measures*"[7]. During the first decade of the application of the Conventions some major conflicts on the interpretation of the notion "pollution damage" have arisen between the 1971 IOPC Fund and certain State Parties, such as Italy and the former USSR, granting compensation for environmental damage according to domestic legislation[8]. The absence of a clear reference to "*environmental damage*" within the 1969 Civil Liability and the 1971 Fund Conventions and the consequential conflicts

---

1971 (27 September 2000, *Ratification of Maritime Conventions*, I.7.112) provided for special arrangements concerning the cession to be in force of the 1971 Fund Convention (*i.e.* on 24 May 2002).

[5] The consolidated texts of the 1969 Civil Liability Convention and the 1971 Fund Convention as amended by the 1992 Protocols are hereinafter referred to as the 1992 Civil Liability Convention and the 1992 Fund Convention.

[6] Resolution on the adoption of amendments of the limitation amounts in the Protocol of 1992 to amend the International Convention on Civil Liability for Oil Pollution Damage, 1969, adopted by the Legal Committee of the International Maritime Organization on 18 October 2000 and Resolution on the adoption of amendments of the limits of compensation in the Protocol of 1992 to amend the International Convention on the Establishment of an International Fund for Compensation for Oil Pollution Damage, 1971, adopted by the Legal Committee of the International Maritime Organization on 18 October 2000.

[7] Article I 1969 Civil Liability Convention; Article 1 1971 Fund Convention.

[8] See more extensive *supra* in this book, NICHOLS, J., *Scope of compensation for environmental damage under the 1992 Civil Liability Convention and the 1992 Fund Convention* (Part II – Chapter 4).

with the provisions as laid down in national legislation[9] have resulted in an explicit amendment of the 1969 Civil Liability and 1971 Fund Conventions through the adoption of the 1992 Protocols, stating that compensation for pollution damage is *inter alia* to be granted for loss or damage caused outside the ship by contamination resulting from the escape or discharge of oil from the ship, wherever such escape or discharge may occur, *provided that compensation for impairment of the environment other than loss of profit from such impairment shall be limited to costs of reasonable measures of reinstatement actually undertaken or to be undertaken*[10]. The limitation of compensation for environmental damage to the costs of measures of reinstatement actually taken or to be undertaken has been recalled in various other international agreements[11] and may partly be explained by the absence of general, scientifically accepted quantification methods to assess the value of the (marine) environment and its components and partly by the scientific uncertainty as to the ability of the environment to recover in a natural way.

1.3. The growing pression of supranational organisations

Although the compensation limits have been sensibly increased in 1992 and 2000, as a way to counter the critics on the inadequacy of the international liability and compensation regime, some European and national fora considered that even the increased limits might be insufficient following catastrophic incidents. Therefore, in the aftermath of the *Erika* incident, the European Commission proposed a Regulation on the establishment of a European Supplementary Fund for Compensation for Oil Pollution Damage (hereinafter referred to as the COPE Fund Proposal) and higher compensation limits for oil pollution damage in European waters. The international regime has reacted strongly against this European proposal with an aim to maintain the uniformity of the international regime and to avoid a second unilateral movement – after the United States adopted their own OPA'90 regime. Accordingly, the proposal initially suggested by the European Commission as a means of providing additional compensation within the European Union was

---

[9] In recent years more and more supranational (*e.g.* Directive 2004/35/EC on environmental liability) and national legislation (*e.g.* Belgium, P.R. China, ex-USSR-countries, Italy, Russia, United Arab Emirates, Venezuela, ...) provides for compensation for impairment of the (marine) environment.

[10] As a consequence of this clarification of the Convention texts, the freedom of interpretation of a domestic judge concerning the extent of "pollution damage" has been strongly restricted. The fact that a domestic judge has no right to further interpret a clear wording within an international convention results from the general rule of interpretation written down in Article 31 of the Vienna Convention on the Law of Treaties. It is indeed striking to ascertain that in the period after the entry into force of the 1992 Protocols, the number of conflicts relating to the admissibility of claims for compensation for environmental damage has been limited and no more out-of-court settlements have been concluded.

[11] See *e.g.* Article 2(2)c of the Protocol of 10 December 1999 on Liability and Compensation for Damage Resulting from Transboundary Movements of Hazardous Wastes and their Disposal to the Basel Convention of 22 March 1989 on the Control of Transboundary Movements of Hazardous Wastes and their Disposal (http://www.basel.int; not yet entered into force) and Article 2(7) of the Convention of 21 June 1993 on Civil Liability for Damage Resulting from Activities Dangerous to the Environment (http://conventions.coe.int; not yet entered into force).

taken up and developed at a global level by the 1992 IOPC Fund and the IMO resulting in the adoption of a third tier of compensation.

This contribution will focus on the impact of EC decision-making on the international regime for oil pollution damage. Firstly, the establishment and the extent of EC competence in the field of maritime transport will be explained and illustrated with European legal initiatives on the subject taken in the aftermath of the *Erika* and *Prestige* incidents. Secondly, the EC COPE Fund Proposal will be examined to demonstrate the influence of this EC proposal on the establishment of a third tier of compensation at the international level. Finally, the 2003 Supplementary Fund Protocol, establishing the third tier of compensation, will be analyzed to clearly show the influences of European legislation and the COPE Fund Proposal.

2. EC COMPETENCE ON MARITIME TRANSPORT

2.1. EC Common Transport Policy

Various references to a community competence in the field of transport are to be detected within the consolidated version of the Treaty establishing the European Community[12] (hereinafter called EC Treaty). Article 3(1)f includes the establishment of a common policy in the sphere of transport[13] in the activities of the

[12] Consolidated version of the Treaty establishing the European Community, *OJ* C 325, 24 December 2002. European integration is based on four founding treaties: the Treaty establishing the European Coal and Steel Community (ECSC), was signed on 18 April 1951 in Paris, entered into force on 23 July 1952 and expired on 23 July 2002. The Treaty establishing the European Economic Community (EEC) and the Treaty establishing the European Atomic Energy Community (Euratom) were signed in Rome on 25 March 1957, and entered into force on 1 January 1958. These Treaties have been signed by the founding Member States Belgium, France, Germany, Italy, Luxemburg and the Netherlands. The Treaty on the European Union was signed in Maastricht on 7 February 1992 and entered into force on 1 November 1993. The "Maastricht Treaty" changed the name of the European Economic Community to simply "the European Community". By introducing new forms of co-operation between the Member State governments – for instance on defence, and in the area of "justice and home affairs" – the Maastricht Treaty created a new structure with three "pillars", the European Union (EU). Moreover, the founding treaties have been amended on several occasions, in particular when new Member States acceded in 1973 (Denmark, Ireland, United Kingdom), 1981 (Greece), 1986 (Spain, Portugal) and 1995 (Austria, Finland, Sweden). There have also been more far-reaching reforms bringing major institutional changes and introducing new areas of responsibility for the European institutions: the Merger Treaty, signed in Brussels on 8 April 1965 and in force since 1 July 1967, provided for a Single Commission and a Single Council of the then three European Communities; the Single European Act (SEA), signed in Luxembourg and the Hague, and entered into force on 1 July 1987, provided for the adaptations required for the achievement of the Internal Market; the Treaty of Amsterdam, signed on 2 October 1997, entered into force on 1 May 1999 *inter alia* amended and renumbered the EU and EC Treaties; the Treaty of Nice, signed on 26 February 2001, that entered into force on 1 February 2003, amended the decision-making system within the European institutions. The EC Treaty, the EU Treaty and the Treaty of Nice have been merged into one consolidated version. Further changes result from the Treaty on the Accession of ten new Member States, signed on 16 April 2003 (entry into force 1 May 2004) and the Treaty establishing a Constitution for Europe, signed in Rome on 29 October 2004 (Source: http://www.europa.eu.int).

[13] The EC Treaty also refers to the establishment and development of trans-European networks in the area of transport as a decisive element in the strengthening of the economic and social cohesion of the

Community for the purposes set out in Article 2, namely: "*The Community shall have as its task, by establishing a common market and an economic and monetary union and by implementing common policies or activities referred to in Articles 3 and 4, to promote throughout the Community a harmonious, balanced and sustainable development of economic activities, a high level of employment and of social protection, equality between men and women, sustainable and non-inflationary growth, a high degree of competitiveness and convergence of economic performance, a high level of protection and improvement of the quality of the environment, the raising of the standard of living and quality of life, and economic and social cohesion and solidarity among Member States*".

Within the EC Treaty the establishment of the common transport policy is determined by the provisions of Title V on transport (ex Title IV)[14]. Article 70 (ex 74) stipulates that the objectives of the EC Treaty shall, in matters governed by title V (ex Title IV), be pursued by Member States within the framework of a common transport policy. Therefore, and taking into account the distinctive features of transport, the Council shall, acting in accordance with the co-decision procedure referred to in Article 251 (ex 189b)[15] and after consulting the Economic and Social Committee (ESC) and the Committee of the Regions (COR), lay down[16]:

European Union (Article 154-156 (ex 129 b-d) EC Treaty). Article 161 (ex 130d) EC Treaty provides for a Cohesion Fund providing a financial contribution to projects in the field of trans-European networks in the area of transport infrastructure. This trans-European networks however are not to be governed by the general provisions relating to transport.

[14] Articles 70-80 (ex 74-84) EC Treaty.

[15] The co-decision procedure of Article 251 (ex 189b) EC Treaty establishes a direct dialogue between the European Parliament and the Council guaranteeing that the adoption of legislation depends on approval by both institutions. Therefore, the Commission submits a proposal to the European Parliament and the Council. The European Parliament formulates an opinion. If the Council approves all the amendments contained in the European Parliament's opinion or if the European Parliament does not propose any amendments, the Council may adopt the proposed (amended) act, acting by a qualified majority. Thus, the Commission cannot act against legislation commonly agreed by the Council and the European Council. In all other cases the Council shall adopt a common position and communicate it to the European Parliament. Hereby the Council shall inform the European Parliament fully of the reasons which led to the adoption of its common position and the European Commission shall inform the European Parliament of its position. If, within three months of such communication, the European Parliament approves the common position or has not taken a decision, the act in question shall be deemed to have been adopted in accordance with that common position. If – contrarily – the European Parliament rejects, by an absolute majority of its component members, the common position, the proposed act shall be deemed not to have been adopted. If the European Parliament proposes amendments to the common position by an absolute majority of its component members, the amended text shall be forwarded to the Council and to the Commission, which shall deliver an opinion on those amendments. If, within three months of the matter being referred to it, the Council, acting by a qualified majority, approves all the amendments of the European Parliament, the act in question shall be deemed to have been adopted in the form of the common position thus amended. However, the Council shall act unanimously on the amendments on which the Commission has delivered a negative opinion. If the Council does not approve all the amendments, the President of the Council, in agreement with the President of the European Parliament, shall within six weeks convene a meeting of the Conciliation Committee. The Conciliation Committee shall have the task of reaching agreement on a joint text, by a qualified majority of the Members of the Council or their representatives and by a majority of the representatives of the European Parliament. The Commission shall take part in the Conciliation Committee's proceedings and shall take all the necessary initiatives with a view to reconciling the positions of the European Parliament and the

(i) common rules applicable to international transport to or from the territory of a Member State or passing across the territory of one or more Member States;
(ii) the conditions under which non-resident carriers may operate transport services within a Member State;
(iii) measures to improve transport safety[17]; and
(iv) any other appropriate provisions.

Articles 72-79 (ex 76-83) EC Treaty fully relate to the abolition of any discrimination on grounds of nationality and of governmental influence on competitive conditions relating to carriers of other Member States compared with carriers who are nationals of that State (including transport rates and conditions)[18].

Council. In fulfilling this task, the Conciliation Committee shall address the common position on the basis of the amendments proposed by the European Parliament. If, within six weeks of its being convened, the Conciliation Committee approves a joint text, the European Parliament, acting by an absolute majority of the votes cast, and the Council, acting by a qualified majority, shall each have a period of six weeks from that approval in which to adopt the act in question in accordance with the joint text. If either of the two institutions fails to approve the proposed act within that period or where the Conciliation Committee does not approve a joint text, the proposed act shall be deemed not to have been adopted (article 251 (ex 189 b) EC Treaty). Article 71(2) (ex 75(2)) EC Treaty) provides for a derogation from the procedure described above where the application of provisions concerning the principles of the regulatory system for transport would be liable to have a serious effect on the standard of living and on employment in certain areas and on the operation of transport facilities. See also LENAERTS, K. & VAN NUFFEL, P. (2003), *Europees recht in hoofdlijnen*, Maklu, Antwerp.

[16] Article 71(1) EC Treaty (ex Article 75(1)).

[17] The provision on measures to improve transport safety was introduced by Article G(16) of the Maastricht Treaty on the European Union (*OJ* C 191, 29 July 1992).

[18] Article 72 (ex 76) contains a transitional provision preventing the adoption of less favourable provisions in their direct or indirect effect on carriers of other Member States as compared with carriers who are nationals of that State. Article 73 (ex 77) contains provisions as to the compatibility of aids with the EC Treaty. Article 74 (ex 78) highlights the principle of taking account of the economic circumstances of carriers in the adoption of any measures in respect of transport rates and conditions. Article 75 (ex 79) relates to the abolishment of any discrimination which takes the form of carriers charging different rates and imposing different conditions for the carriage of the same goods over the same transport links on grounds of the country of origin or of destination of the goods in question in the case of transport within the Community. Article 76 (ex 80) prohibits the imposition by a Member State - unless authorized by the Commission - in respect of transport operations carried out within the Community, of rates and conditions involving any element of support or protection in the interest of one or more particular undertakings or industries and lays down the procedure for authorization. This prohibition however does not apply to tariffs fixed to meet competition. Article 77 (ex 81) relates to charges or dues in respect of the crossing of frontiers which are charged by a carrier in addition to the transport rates. Article 78 (ex 82) contains a specific provision for the Federal Republic of Germany in order to take into account the economic disadvantages caused by the division of Germany. Article 79 (ex 83) attaches an Advisory Committee to the Commission and states that the Commission, whenever it considers it desirable, shall consult the Committee on transport matters without prejudice to the powers of the Economic and Social Committee.

2.2. EC competence on maritime transport

2.2.1. The extent of article 80 (ex 84) EC Treaty
Although the competence of the European Union in transport may thus appear clearly described in the EC Treaty, the last article of Title V contains divergent provisions as to its application on maritime transport matters: Article 80(1) (ex 84(1)) EC Treaty stipulates that the abovementioned provisions of Title V on transport shall apply only to transport by rail, road and inland waterways. Nevertheless, Article 80(2) (ex 84(2)) EC Treaty adds that the Council, *acting by a qualified majority*, may decide whether, to what extent and by what procedure appropriate provisions may be laid down for sea and air transport, *applying the procedural provisions of Article 71 (ex 75)*.
Before the modifications introduced by Article 6 and 16(5) Single European Act (SEA - 1986)[19], this decision was to be taken *unanimously*, thus putting a heavy burden on the development of a common maritime transport policy. However, the amendments have been carelessly drafted: the first paragraph still refers to "*by what procedure*", although the last part of it explicitly refers to the procedure as laid down in Article 71 (ex 75). Furthermore the words "*whether*" and "*to what extent*" have not been deleted, thus still evoking questions on the obligatory character of the establishment of a common policy on maritime (and air) transport[20].
Article III-245 of the Treaty establishing a Constitution for Europe[21] does not seem to clarify the reasons as to why a divergent regime has been installed and sticks to the sole application of Section 7 on Transport to transport by rail, road and inland waterways (Article III-245(1)). However, the procedural modifications seem to clear the way to an outlined EC maritime transport competence: *"European laws or framework laws may lay down appropriate measures for sea and air transport. They shall be adopted after consultation of the Committee of the Regions and the Economic and Social Committee"* (Article III-245(2)).

2.2.2. Explanations
FARANTOURIS resumes the explanations, relating in particular to the special characteristics of the maritime transport sector, as to why the founders of the European Community viewed maritime transport as different from other types of transport and exempted it from normal application within the Common Transport Policy, as follows: "*(...) certain special features which make the maritime transport market different from other sectors of the economy and, as clarified below, are often not freely compatible with the fundamental principles of the Common market, such as the abolition of any discrimination on grounds of nationality and of governmental influence on competitive conditions. Among these pecularities is the political sensitivity within the area which has led to high degree of government involvement, including market regulation and public funding. It is also characteristic that there is*

[19] *O.J.* L 169/1, 29 June 1987.
[20] KAPTEYN, P. & VERLOREN VAN THEMAAT, P. (2003), *Het recht van de Europese Unie en van de Europese* Gemeenschappen, Kluwer, Deventer, 1.008.
[21] Treaty establishing a Constitution for Europe, signed in Rome on 29 October 2004 *O.J.* C 310, 16 December 2004, not yet in force.

a differentiation in the treatment of the means of maritime transport (tramp carriers, liner maritime transport services, ferries etc.) and the infrastructure which they use. The infrastructure is often in different ownership and subject to public intervention. There is also a very strong international element in maritime transport, and there are many bilateral and international agreements to which most of the Member States had become signatories prior to the creation of the European Community. Other examples of the special though not necessarily unique aspects of sea transport are: the fact that competition may take place within modes and different modes; the desirability of co-operation between different modes; technical and economic barriers to access to the market (e.g. in the case of liner conferences and liner consortia); the great mobility of capital assets of the maritime transport sector which distinguishes it from inland transport and the manufacturing industry; specialised markets and specialised technology; the environmental impact of maritime transport etc."[22].

This reference to the special characteristics of the maritime transport sector has been recalled by Advocate General REISCHL in the Case 167/73 *Commission v. France*: "*In recognition of the fact that special problems attach to maritime transport, there is a special provision in Article 84* (now Article 81) *that the provisions on transport do not apply*"[23].

Although this may very well explain the *prima facie* exclusion of maritime transport from the common transport policy, the paragraph was in fact inserted due to the wishes of the Netherlands which sought to have very little (if any) involvement by the EC in shipping matters[24].

2.2.3. Consequences

The question arises as to what consequences must be attached to the exemption installed by Article 80 (ex 84) EC Treaty: is maritime transport only to be exluded from the application of the common transport policy provisions or also from the application of the general rules of the EC Treaty, in particular free movement of goods, persons, services and capitals and the rules of competition? Article 51(1) (ex 61(1)) EC Treaty, determining that freedom to provide services in the field of transport shall be governed by the provisions of the title relating to transport, might tempt us to conclude *a contrario* that the other provisions of the EC Treaty do not apply to the maritime transport sector. The European Court of Justice (ECJ) has decided otherwise. In the aforementioned case 167/73 *Commission v. France*, the so-called *French Merchant Seamen Case*, the Court ruled that "*far from excluding the application of the basic principles of the EEC Treaty to sea and air transport, Article 84(1)* (now Article 80(1)) *provides only that the special provisions of the title relating to transport shall not automatically apply to them and that therefore sea*

[22] FARANTOURIS, N.E. (2003), *European integration & maritime transport*, Ant. N. Sakkoulas/Bruylant, Athens/Brussels, 89-92.

[23] ECJ 4 April 1974, *Commission/France*, C-167/73, [1974] *ECR* 359; http://www.curia.eu.int.

[24] Working Documents; Doc. 5/77; 23 March 1977; Rapporteur Mr. H. SEEFIELD, paras 1 and 3 mentioned by POWER, V. (1992), *EC Shipping Law*, Lloyd's of London Press Ltd , London, 110.

and air transport remained, like other modes of transport, subject to the general rules of the Treaty. It thus follows that the application of the basic principles to the sphere of sea transport is not optional but obligatory for Member States. Since transport is basically a service, it has been found necessary to provide a special system for it, taking into account the special aspects of this branch of activity. With this object, a special exemption has been provided by Article 61(1) (now Article 51(1)), *under which freedom to provide services in the field of transport 'shall be governed by the provisions of the title relating to transport', thus confirming that the general rules of the Treaty must be applied insofar as they are not excluded. The basic rules have been conceived as being applicable to the whole complex of economic activities and can be rendered inapplicable only as a result of express provision in the Treaty. The provisions of Title IV* (now Title V) *on Transport however do not contain such an exemption. When Article 74* (now Article 70) *refers to the objectives of the Treaty, it means the provisions of Articles 2 and 3, for the attainment of which the fundamental provisions applicable to the whole complex of economic activity are of prime importance*[25].

In the *Nouvelles Frontières Case* the Tribunal de Police de Paris referred a question for a preliminary ruling to the Court in order to enable it to appraise the compatibility of the provisions of the compulsory approval procedure laid down by French law for air tariffs with the competition rules in the EEC treaty. The European Court of Justice relied in this respect on the aforementioned judgment of 4 april 1974. The French Government submitted that the solution propounded by the Court in the aforementioned judgment was referring only to the rules contained in part II of the Treaty on the foundations of the community and could therefore not be transposed to the competition rules which are contained in part III of the Treaty concerning the policy of the Community. Nevertheless, the Court ruled that: "*as regards transport - there is no provision in the Treaty which excludes the application of the competition rules or makes it subject to a decision by the Council. It follows that air transport remains, on the same basis as the other modes of transport, subject to the general rules of the Treaty, including the competition rules*"[26].

---

[25] *Ibid.* According to the Ministerial Order issued in implementation of article 3(2) of the French Code du Travail Maritime (French Merchant Seamen's Code), the major proportion of the crew of French merchant vessels were required to be French nationals and key positions on board were only to be taken by French nationals. In 1996 the French government has once again been convicted for not having amended the alleged provision by means of national provisions of a binding nature which have the same legal force as those which must be amended: "*Mere administrative practices, which by their nature are alterable at will by the authorities and are not given the appropriate publicity, cannot be regarded as constituting the proper fulfilment of obligations under the Treaty*" (ECJ 7 March 1996, *Commission/France,* C-334/94, [1996] *ECR* I-01307; http://www.curia.eu.int).

[26] ECJ 30 April 1984, *Ministère Public/Lucas Asjès et al. (Nouvelles Frontières),* C-209-213/84, [1986] *ECR* 1425; http://www.curia.eu.int.

Following these landmark judgments, a whole range of other cases have followed, clarifying the application of the freedoms of movement of goods, persons, services and capitals and the rules of competition on the maritime transport sector[27].

2.3. Exercise of EC competence on maritime transport

2.3.1. The development of a Common Maritime Transport Policy

Although the establishment of a common transport policy had already been called upon in the founding Treaty of the EEC, the development thereof has been very slow. Following a barren period in the 1960's[28], the first enlargements of the EEC in 1973 (with the accession of the UK, Ireland and Denmark) and in 1981 (with the accession of Greece), the adoption of the UN Convention on a Code of Conduct for Liner Conferences[29] and the occurrence of several oil tanker disasters have led to the non-cohesive, sporadic adoption of (maritime) transport legislation[30]. Therefore, in 1983 the European Parliament, hereby supported by the European Commission, instituted proceedings in the ECJ against the Council of Ministers for failing to introduce a common policy for transport in a binding manner — in particular for failing to ensure freedom to provide services in the sphere of international transport and to lay down the conditions under which non-resident carriers may operate transport services in a Member State — and further for failing to reach a decision on 16 specified proposals submitted by the European Commission in relation to transport[31].

Following this landmark judgment, the European Commission published three memoranda on *Progress towards a Common Transport Policy*, in particular *Progress towards a Common Maritime Policy: Maritime Transport*, including a package of proposals for legislation (14 March 1985)[32]. This policy paper

[27] For an overview, see FARANTOURIS, N.E. (2003), *o.c.*, 465 p.; POWER, V. (1992), *o.c.*, 771 p.

[28] *E.g.* on 26 November 1962 the Council of Ministers adopted Regulation 141/62 exempting transport from the application of Regulation 17 (*OJ* P 124, 28 November 1962) to temporarily non-apply Articles 85-94 EC Treaty relating to the competition rules on sea and air transport. This Regulation has not been repealed until the adoption of Regulation 4056/86.

[29] UN Convention on a Code of Conduct for Liner Conferences (Geneva, 6 April 1974, entry into force on 6 October 1983), *UNTS*, Vol. 1334, 15 and Vol. 1365, 360.

[30] *E.g.* Directive 79/115/EEC of 21 December 1978 concerning pilotage of vessels by deep-sea pilots in the North Sea and English Channel, *OJ* L 33, 8 Februay 1979. For a complete overview, see POWER, V. (1992), *o.c.*, 133-158 and PALLIS, A.A. (2002), *The Common EU Maritime Transport Policy – Policy Europeanisation in the 1990s*, Ashgate Publishing Ltd., Hampshire, 58-70.

[31] ECJ 22 May 1985, *European Parliament v. Council of the European Communities*, C-13/83, [1985] *ECR* 1.513; http://www.curia.eu.int. See also STEVENS, H. (2004), *Transport Policy in the European Union*, The European Union Series (Palgrave MacMillan), Hampshire/New York, 47-65.

[32] EUROPEAN COMMISSION, *Progress towards a Common Maritime Policy: Maritime Transport*, COM (1985) 90 final, 14 March 1985, *OJ* C 212, 23 August 1985; The European Commission proposed for a Draft Council Regulation concerning Co-Ordinated Action to Safeguard Free Access to Cargoes in Ocean Trades; a Draft Council Regulation applying the Principle of Freedom to Provide Services to Sea Transport; a Draft Council Regulation on Unfair Pricing Pratices in Maritime Transport; a Draft Council Decision amending Decision 77/587/EEC setting up a Consultation Procedure on Relations between

constituted the first comprehensive action on maritime transport. Furthermore, the *White Paper on Competing the Internal Market* referred to the necessity of making rapid progress in the transport area and set a series of targets between 1986 and 1989 for the adoption of single market measures by the Council in relation to all transport modes except rail[33]. Following the facilitation of the enactment of maritime transport legislation through the amendments of the SEA (see above) a second comprehensive action has been undertaken by the European Commission with the adoption of *A Future for the Community Shipping Industry: Measures to Improve the Operating Conditions of Community Shipping* (3 August 1989)[34].

Since 1992 the European Commission has continued to develop its transport policies on a pragmatic modal basis with a strong emphasis on measures required to buttress the single market. In addition, under the Treaty of Maastricht, the European Commission had acquired new powers for transport safety and for transport infrastructure, and its influence as the principal spokesman and negotiator for the EC in international bodies dealing with transport policies, especially within Europe, was continuing to grow. At the same time the EC Treaty has been amended to lay increasing emphasis on environmental protection: Article 130r (now 174) gave the environment a rather special status under the EC Treaty by requiring that "*environmental protection requirements must be integrated into the definition and implementation of other Community policies*"[35]. Moreover, the *White Paper on The Future Development of the Common Transport Policy* carried the subtitle *A Global Approach to the Construction of a Community Framework for Sustainable Mobility*, signalling the intention to give more emphasis to environmental issues in determining the content of the Common Transport Policy[36].

Member States and Third Countries in Shipping Matters and on Action Relating to such Matters in International Organizations; a Draft Council Directive concerning a Common Interpretation of the Concept of "National Shipping Line" and finally, amendments to the Proposal for a Council Regulation laying down Detailed Rules for the Application of Articles 85 and 86 of the Treaty to Maritime Transport.

[33] STEVENS, H. (2004), *o.c.*, 59.

[34] EUROPEAN COMMISSION, *A Future for the Community Shipping Industry: Measures to Improve the Operating Conditions of Community Shipping*, COM (1989) 266 final, 3 August 1989, *OJ* C 263, 16 October 1989, including Proposal for a Council Regulation Establishing a Community Ship Register and Providing for the Flying of the Community Flag by Sea-Going Vessels; Proposal for a Council Regulation on a Common Definition of a Community Shipowner; and Proposal for a Council Regulation applying the Principle of Freedom to Provide Services to Maritime Transport within Member States.

[35] In recent years, the European policies including the maritime transport policy, have been reorientated towards sustainable development. For an overview, see GONSAELES., G. (2002), The operationalisation of sustainable development in the European policy on maritime transport, in: BARUDOV, S.T., BOGDANOV, P.A. & DRAGANCHEV, H.S. (eds.), *Proceedings of the First International Congress on Mechanical and Electrical Engineering and Technology (MEET '2002) and the Fourth International Conference on Marine Industry (MARIND '2002)*, Varna, 6-11 oktober 2002, 219-226.

[36] STEVENS, H. (2004), *o.c.*, 61-62.

2.3.2. The development of the EC Maritime Safety Policy

It is against this background that the EC policy on maritime safety has been developed[37]. The EC Maritime Safety Policy is based on the Communication from the European Commision *A common policy on safe seas* (1993), analyzing the maritime safety situation in Europe and outlining a framework for a Common Maritime Safety Policy based on four pillars: (i) convergent implementation of existing international rules; (ii) uniform enforcement of international rules by port states; (iii) development of navigational aids and traffic surveillance infrastructures, and; (iv) reinforcement of the role of the EU as driving force for international rule-making. An action programme attached to this Communication highlighted the main decisions to be taken to improve maritime safety in Europe and to better protect the European coasts[38].

As a consequence, the main legislation adopted in this period aimed at ensuring the implementation of international safety rules by all ships visiting European ports and at ensuring that ships flying a flag of a EU Member State and their crew comply with the international standards. This legislation relates to port state control[39], classification societies[40], the human element factor[41], minimum requirements for ships carrying dangerous or polluting goods[42], tonnage measurement of ballast

---

[37] See more extensive, PALLIS, A.A. (2002), *o.c.*, 103-166.

[38] EUROPEAN COMMISSION, *A common policy on safe seas*, COM (1993) 66 final, 24 February 1993; http://europa.eu.int/comm/transport/maritime/safety/1993_en.htm.

[39] Directive 95/21/EC of 19 June 1995 concerning the enforcement, in respect of shipping using Community ports and sailing in the waters under the jurisdiction of the Member States, of international standards for ship safety, pollution prevention and shipboard living and working conditions (port State control) (*OJ* L 157, 7 July 1995), as amended by Directive 98/25/EC of 27 April 1998 (*OJ* L 133, 7 May 1998), by Directive 98/42/EC of 19 June 1998 (*OJ* L 184, 27 June 1998); by Directive 1999/97/EC of 13 December 1999 (*OJ* L 331, 23 December 1999); by Directive 2001/106/EC of 19 December 2001 (*OJ* L 19, 22 January 2002); and by Directive 2002/84/EC of 5 November 2002 (*OJ* L 324, 29 November 2002).

[40] Directive 94/57/EC of 21 November 1994 on common rules and standards for ship inspection and survey organizations and relevant activities of maritime administrations (*OJ* L 319, 12 December1994) as amended by Directive 97/58/EC of 26 September 1997 (*OJ* L 274, 7 October 1997); and by Directive 2001/105/EC of 19 December 2001 (*OJ* L 19, 22 January 2002).

[41] Directive 94/58/EC of 22 November 1994 on the minimum level of training of seafarers (*OJ* L 319, 12 December 1994), as amended by Directive 2001/25/EC of 4 April 2001 (*OJ* L 136, 18 May 2001); and by Directive 2003/103/EC of 17 November 2003 (*OJ* L 326, 13 December 2003). See also Directive 1999/63/EC of 21 June 1999 concerning the Agreement on the organization of working time of seafarers concluded by the European Community Shipowners' Association (ECSA) and the Federation of Transport Workers' Unions in the European Union (FST) (*OJ* L 167, 2 July 1999) and Directive 1999/95/EC of 13 December 1999 concerning the enforcement of provisions in respect of seafarers' hours of work on board ships calling at Community ports (*OJ* L 14, 20 January 2000).

[42] Directive 93/75/EEC of 13 September 1993 concerning minimum requirements for vessels bound for or leaving Community ports and carrying dangerous or polluting goods (*OJ* L 247, 5 October 1993), as amended by Directive 97/34/EC of 6 June 1997 (*OJ* L 158, 17 June 1997); by Directive 98/55/EC of 17 July 1998 (*OJ* L 215, 1 Augustus 1998); and by Directive 98/74/EC of 1 October 1998 (*OJ* L 276, 13 October 1998) and repealed by Directive 2002/59/EC of 27 June 2002 establishing a Community vessel traffic monitoring and information system and repealing Council Directive 93/75/EEC (*OJ* L 208, 5 Augustus 2002).

spaces in segregated ballast oil tankers[43], marine equipment[44] and a harmonized safety regime for fishing vessels[45].

Some major incidents with passenger ships – such as the *Estonia*, which sunk on 28 September 1994 in the Baltic sea – resulted in the adoption of new measures in order to ensure:

(i) harmonized safety standards for all new and existing passenger vessels and high speed craft[46];
(ii) the establishment of a system of mandatory surveys for passenger ships and high-speed crafts operating on regular international or domestic service to or from EU ports[47];
(iii) an improved registration of persons on board passenger's ships[48]; and
(iv) an early implementation of the International Safety Management Code (ISM Code) on roll-on/roll-of passenger ferries[49].

In its *Communication on the enhanced safety of passenger ships in the Community*[50], the European Commission proposed for a set of measures aiming at improving existing EC legislation on the safety of passenger ships, in line with the policy objectives outlined in the *White Paper on Transport Policy*[51].

2.3.3. The European Transport Policy 2001-2010

The *White Paper "European Transport Policy for 2010: time to decide"*, emphasizing the need for the optimization of the transport system to meet the demands of enlargement and sustainable development, was approved on 12

---

[43] Regulation 2978/94 of 21 November 1994 on the implementation of IMO Resolution A.747(18) on the application of tonnage measurement of ballast spaces in segregated ballast oil tankers (*OJ* L 319 of 12.12.1994), to be repealed as from 31 December 2007 as a result of Regulation 417/2002 of 18 January 2002 (*OJ* L 64, 7 March 2002).

[44] Directive 96/98/EC of 20 December 1996 on marine equipment (*OJ* L 046, 17 February 1997), as amended by Directive 98/85/EC of 11 November 1998 (*OJ* L 315, 25 November 1998).

[45] Directive 97/70/EC of 11 December 1997 setting up a harmonised safety regime for fishing vessels of 24 metres in length and over (*OJ* L 034, 9 February 1998).

[46] Directive 98/18/EC of 17 March 1998 on safety rules and standards for passenger ships (*OJ* L 144, 15 May 1998).

[47] Directive 1999/35/EC of 29 April 1999 on a system of mandatory surveys for the safe operation of regular ro-ro ferry and high-speed passenger craft services (*OJ* L 138, 1 June 1999).

[48] Directive 98/41/EC of 18 June 1998 on the registration of persons sailing on board passenger ships operating to or from ports of the Member States of the Community (*OJ* L 188, 2 July 1998).

[49] Regulation 3051/95 of 8 December 1995 on the safety management of roll-on/roll-off passenger ferries (ro-ro ferries) (*OJ* L 320, 30 December 1995).

[50] EUROPEAN COMMISSION, *Communication from the Commission on the enhanced safety of passenger ships in the Community*, COM (2002) 158 final, http://europa.eu.int/eur-lex/en/com/pdf/2002/com2002_0158en01.pdf.

[51] These measures have been adopted on 14 April 2003 (Directive 2003/24/EC of the European Parliament and of the Council of 14 April 2003 amending Council Directive 98/18/EC on safety rules and standards for passenger ships (*OJ* L 123, 17 May 2003) and Directive 2003/25/EC of the European Parliament and of the Council of 14 April 2003 on specific stability requirements for ro-ro passenger ships (*OJ* L 123, 17 May 2003)).

September 2001[52]. This *White Paper* sought an answer for the unequal growth in the different modes of transport, the congestion on the main routes and the harmful effects on the environment and public health through a series of measures ranging from pricing to revitalizing alternative modes of road transport and targeted investment in the trans-European network (TEN)[53].

The *White Paper* also pointed out that the EU should have tougher rules on maritime safety going beyond those proposed in the aftermath of the *Erika* disaster (see below). In setting out its maritime safety policy for the coming decade, the extension of the EC seaboard upon enlargement has been taken into account. The *White Paper* expects that enlargement should enable more stringent controls to be carried out on ships in all ports, which should lead to the gradual disappearance of ports of convenience from the European continent. Enlargement should also be the occasion to include not only technical requirements regarding ships' structure and maintenance in the criteria to be met by ships calling at European ports, but also social standards, starting with the standards for seafarers of the International Labour Organization (ILO). To this end, the European Commission should define the minimum social conditions it intends to enforce for crews[54].

Moreover, the objectives set out in the *White Paper* aimed at organizing the monitoring of shipping more effectively and minimizing the risk of accidents, particularly those caused by ships carrying dangerous or polluting goods. Therefore, "*the lack of any powers of inspection or enforcement on the part of the IMO*" should be tackled. To combat ports and flags of convenience more effectively, the European Commission — in collaboration with the IMO and the ILO — should propose measures incorporating the minimum social rules to be observed in ship inspections and developing a genuine european maritime traffic management system. By 2010, the enlarged Union should, as in the air transport sector, have a traffic management system in place to protect itself against dangerous or suspicious movements of ships, in particular by diverting them to ports of refuge. In addition, best social or fiscal practices developed at national level, such as the tonnage-based taxation system, should be emulated to promote the reflagging of as many ships as possible to Community registers. Under this system, shipowners pay a tax based on the tonnage they operate, regardless of the actual earnings of the business. Targeted checks on certain flags of convenience need to be combined with measures taken in the framework of the European policy on judicial cooperation.

---

[52] EUROPEAN COMMISSION, *White Paper: European Transport Policy for 2010: Time to decide*,

[53] The Gothenburg European Council placed shifting the balance between modes of transport at the heart of the sustainable development strategy. This implies *inter alia* a shift of balance between the modes by way of an investment policy in infrastructure geared to shortsea shipping and the inland waterways (EUROPEAN COUNCIL, *Gothenburg European Council (15-16 June 2001): Presidency conclusions*, http://www.europa.eu.int).

[54] See also EUROPEAN COMMISSION, *Communication from the Commission to the Council and the European Parliament on seafarer training and recruitment*, COM (2001) 188 final, 6 April 2001.

According to the *White Paper*, the European Commission should - at the same time to promote the reflagging of as many ships as possible to the Community register - examine the revision of the guidelines on State aid to maritime transport. The European Commission also planned to propose a framework directive to establish the principles of infrastructure charging and a pricing structure for all modes of transport. Maritime transport should need to integrate charges which incorporate costs relating to maritime safety (especially assistance to shipping at sea, buoyage, availibility of tugs) to be paid by all ships sailing in European waters.

One of the main aims set out in the *White Paper* is to increase the ability of the EC "*to assert itself in the international arena and speak with a single voice in defence of its social, industrial and environmental interests*". Therefore, at the end of 2001 the European Commission proposed that the Council opened negotiations with these organizations with a view to the European Union becoming a full member.

2.3.4. Enforced measures in the aftermath of the Erika and the Prestige

The incidents with the oil tanker *Erika* on 12 December 1999 and the oil tanker *Prestige* on 13 November 2002 have given a fresh impulse for developments concerning maritime safety within the European Union. Only three months after the *Erika* incident, the European Commission presented a first set of measures on maritime safety that were contained in the *Communication on the safety of the seaborne oil trade*, generally known as the *Erika I-package*[55]. Hereby, the European Commission pointed out that "*the normal framework for international action on maritime safety under the auspices of the IMO falls short of what is needed to tackle the causes of such disasters effectively*". The *Erika I-package* proposed for (i) an amended Directive on port state control, (ii) an amended Directive on common rules and standards for ship inspection and survey organizations and for the relevant activities of maritime administrations and (iii) a Regulation on the accelerated phasing-in of double hull or equivalent design requirements for single hull oil tankers.

The *Erika I-package* was completed with the *Communicaton on a second set of Community measures on maritime safety following the sinking of the oil tanker*

---

[55] EUROPEAN COMMISSION, *Communication from the Commission to the European Parliament and the Council on the safety of the seaborne oil trade*, 21 March 2000, COM (2000) 142 def, including *Proposal for a Directive of the European Parliament and of the Council amending Council Directive 95/21/EC concerning the enforcement, in respect of shipping using Community ports and sailing in the waters under the jurisdiction of the Member States, of international standards for ship safety, pollution prevention and shipboard living and working conditions (port State control)*, 21 March 2000, COD (2000) 0065; *Proposal for a Directive of the European Parliament and of the Council amending Council Directive 94/57/EC on common rules and standards for ship inspection and survey organizations and for the relevant activities of maritime administrations*, 21 March 2000, COD (2000) 0066; and *Proposal for a Regulation of the European Parliament and of the Council on the accelerated phasing-in of double hull or equivalent design requirements for single hull oil tankers*, 21 March 2000, COD (2000) 0067.

Erika (*the Erika II-package*)[56]. This communication contained (i) a proposal for a Directive on the establishment of a Community monitoring, control and information system for maritime traffic, (ii) a proposal for a Regulation on the establishment of a European Maritime Safety Agency and (iii) the abovementioned proposal for a Regulation on the establishment of a Fund for the Compensation of Oil Pollution Damage in European Waters and related measures (COPE Fund Proposal).

Following the incident with the *Prestige* the European Commission launched a new *Communication On Improving Safety At Sea In Response to the Prestige Accident*[57]. Within this communication the European Commission stressed the need for the Member States to apply the measures adopted in the *Erika I-* and *Erika II-packages* earlier than the scheduled dates and proposed for a number of additional measures[58]. This communication has been furtherly detailed in the *Report to the European Council on action to deal with the effects of the Prestige disaster*[59].

The proposals laid down in the *Erika I-* and *the Erika II-Packages* have resulted in the adoption of a whole set of new or strenghtened European maritime safety regulations:

(i) Directive 2001/106/EC concerning the enforcement, in respect of shipping using Community ports and sailing in the waters under the jurisdiction of the Member States, of international standards for ship safety, pollution prevention and shipboard living and working conditions (Port State Control) strengtened the existing Directive on Port State control and has been furtherly amended in the aftermath of the *Prestige* indicent by Directive 2002/84/EC amending the Directives on maritime safety and the prevention of pollution from ships[60, 61];

---

[56] EUROPEAN COMMISSION, *Communication from the Commission to the European Parliament and the Council on a second set of community measures on maritime safety following the sinking of the oil tanker Erika*, 6 December 2000, COM (2000) 802def. including *Proposal for a Directive of the European Parliament and of the Council establishing a community monitoring, control and information system for maritime traffic*, 6 December 2000, COD (2000) 0325; *Proposal for a Regulation of the European Parliament and of the Council establishing a European Maritime Safety Agency*, 6 December 2000, COD (2000) 0327; and *Proposal for a Regulation of the European Parliament and of the Council on the establishment of a Fund for the compensation of oil pollution damage in European waters and related measures*, 6 December 2000, COD (2000) 0326.

[57] EUROPEAN COMMISSION, *Communication from the Commission to the European Parliament and to the Council on Improving Safety At Sea in Response to the Prestige Accident*, 3 December 2002, COM (2002) 681 def.

[58] A Proposal for a Directive on ship-source pollution and on the introduction of sanctions, including criminal sanctions, for pollution offences is still pending (Proposal for a Directive of the European Parliament and of the Council on ship-source pollution and on the introduction of sanctions, including criminal sanctions, for pollution offences, COM (2003) 92 final).

[59] EUROPEAN COMMISSION, *Report to the European Council on action to deal with the effects of the Prestige disaster*, 5 March 2003, COM (2003) 105 final.

[60] Directive 2001/106/EC of the European Parliament and of the Council of 19 December 2001 amending Council Directive 95/21/EC concerning the enforcement, in respect of shipping using Community ports and sailing in the waters under the jurisdiction of the Member States, of international standards for ship safety, pollution prevention and shipboard living and working conditions (port State control) (*OJ* L 019,

(ii) Directive 94/57/EC on common rules and standards for ship inspection and survey organizations and relevant activities of maritime administrations has been amended by Directive 2001/105/EC raising the quality requirements for classification societies, subject to a preliminary authorization to operate within the EU. The performance of classification societies will also strictly be monitored, and failure to meet the standards may result in a temporary or permanent withdrawal of their authorization[62];

(iii) Regulation 417/2002 on the accelerated phasing-in of double hull or equivalent design requirements for single hull oil tankers and repealing Regulation 2978/94 contained a timetable for the phasing out of single hull tankers[63]. This timetable has been strengthened following the *Prestige* incident by Regulation 1726/2003[64].

(iv) Directive 2002/59/EC establishing a Community vessel traffic monitoring and information system contains provisions to strengthen the Community vessel traffic monitoring and information system,

---

22 January 2002) as amended by Directive 2002/84/EC of the European Parliament and of the Council of 5 November 2002 amending the Directives on maritime safety and the prevention of pollution from ships (*OJ* L 324, 29 November 2002).

[61] It is worth mentioning that the adoption of the 25th amendment to the Paris MOU (Memorandum of Understanding on Port State Control in Implementing Agreements on Maritime Safety and Protection of the Marine Environment (Paris, 26 January 1982, entry into force: 22 July 2003, *Ratification of Maritime Conventions*, II, 100), brings the Paris MOU in line with the latest changes of the EC Directive on Port State Control, thus extending the geographical scope of application of the EU Directive to *all* the Member States of the Paris MOU (*i.e.* Belgium, Canada, Croatia, Denmark, Finland, France, Germany, Greece, Iceland, Ireland, Italy, Netherlands, Norway, Poland, Portugal, Russian Federation, Slovenia, Spain, Sweden and the UK).

[62] Directive 2001/105/EC of the European Parliament and of the Council of 19 December 2001 (*OJ* L 19, 22 January 2002) amending Directive 94/57/EC of 21 November 1994 on common rules and standards for ship inspection and survey organizations and relevant activities of maritime administrations (*OJ* L 319, 12 December 1994), as formerly amended by Directive 97/58/EC of 26 September 1997 (*OJ* L 274, 7 October 1997).

[63] This timetable was different from the original proposal of the European Commission as a result of the decision of the Council to negotiate within the IMO with a view to amending the International Convention for the Prevention of Pollution from Ships, 1973, as modified by the Protocol of 1978 relating thereto (MARPOL 73/78). The international "double hull" requirements had been introduced into the MARPOL 73/78 Convention in the light of the adoption of OPA'90 by the United States, unilaterally imposing double hull requirements in the form of age limits and thus distorting the - by itself very sensitive - international economic field of maritime transport (amendments adopted on 6 March 1992, entry into force: 6 July 1993). The amendments of 27 April 2001 to Annex I of MARPOL 73/78 (entry into force: 1 September 2002) brought in a new global timetable for accelerating the phase-out of single-hull oil tankers. Under the amendments of 4 December 2003 (entry into force April 2005) the revised regulation 13G of Annex I of MARPOL has brought forward the final phasing-out date for Category 1 tankers (pre-MARPOL tankers) to 2005, from 2007. The final phasing-out date for category 2 and 3 tankers (MARPOL tankers and smaller tankers) has been brought forward to 2010, from 2015. This revisions may be seen as another example of the influence of the EC on the international decision-making process.

[64] Regulation 417/2002 of the European Parliament and of the Council of 18 February 2002 on the accelerated phasing-in of double hull or equivalent design requirements for single hull oil tankers and repealing Council Regulation 2978/94 (*OJ* L 64, 7 March 2002), as amended by Regulation 1726/2003 of the European Parliament and of the Council of 22 July 2003 (*OJ* L 249, 1 October 2003).

amongst others requiring ships built on or after 1 July 2002 and calling at a port of a Member State to be fitted with an automatic identification system (AIS), as well as a voyage data recorder system (VDR, the so-called "*black box*") to facilitate investigations following accidents. This Directive also obliged Member States to draw up plans to accommodate ships in distress in waters under their jurisdiction ("*places of refuge*") by 5 Februari 2004, taking into account the relevant IMO guidelines[65].

(v) Regulation 1406/2002 established a European Maritime Safety Agency (EMSA) to provide the Member States and the European Commission with technical and scientific assistance to ensure the proper application of Community legislation in the field of maritime safety, to monitor its implementation and to evaluate its effectiveness. The tasks of the EMSA has been extended by Regulations 1644/2003 and 724/2004[66].

All proposals as contained in the *Erika I-* and *the Erika II-packages* save the COPE Fund Proposal have thus been adopted, some upon amendment as a consequence of the *Prestige* incident.

3. THE IMPACT OF EC DECISION-MAKING ON THE INTERNATIONAL REGIME FOR OIL POLLUTION DAMAGE

3.1. Preliminary assessment of the international regime

The COPE Fund proposal has been based on the results of an assessment of the international liability and compensation regime for oil pollution damage that was executed by the European Commission. Three criteria had been established against which the adequacy of the international regime needed to be assessed: (i) it should provide prompt compensation to victims without having to rely on extensive and lengthy judicial procedures; (ii) the maximum compensation limit should be set at a sufficiently high level to cover claims from any foreseeable disaster occurring as a result of an oil tanker accident; and (iii) the regime should contribute to discouraging tanker operators and cargo interests from transporting oil in anything other than tankers of an impeccable quality. According to the conclusions of the

[65] Directive 2002/59/EC of the European Parliament and of the Council of 27 June 2002 establishing a Community vessel traffic monitoring and information system and repealing Council Directive 93/75/EEC (*OJ* L 208, 5 Augustus 2002).

[66] Regulation 1406/2002 of the European Parliament and of the Council of 27 June 2002 establishing a European Maritime Safety Agency (*OJ* L 208, 5 Augustus 2002), as amended by Regulation 1644/2003 of the European Parliament and of the Council of 22 July 2003 (*OJ* L 245, 29 September 2003) and by Regulation 724/2004 of the European Parliament and of the Council of 31 March 2004 (*OJ* L 129, 29 April 2004).

European Commission "*the international system satisfied some of these concerns but not all of them*"[67].

3.2. Initial COPE Fund Proposal

The COPE Fund proposal aimed at ensuring adequate compensation for oil pollution damage in European waters by complementing the existing international liability and compensation regime at Community level. This should be realized through the establishment of a European Fund for Compensation for Oil Pollution in European Waters (hereinafter called the COPE Fund) under which higher compensation limits should be available. Furthermore, the proposal aimed at introducing a financial penalty to be imposed on any person who has been found to have contributed to an oil pollution incident by his wrongful intentional or grossly negligent acts or omissions (Article 1).

The proposal followed the geographical scope of application of the 1992 Civil Liability and Fund Conventions (Article 2) and recalled the definitions of "*oil*", "*contributing oil*", "*ton*", "*terminal installation*" and "*incident*" as laid down in these Conventions (Article 3).

Compensation from the COPE Fund would follow the principles and rules of the international compensation regime: the COPE Fund should pay compensation to any person who is entitled to compensation for pollution damage under the 1992 Fund Convention but who has been unable to obtain full and adequate compensation under that Convention, because the totality of valid claims exceed the amount of compensation available under the 1992 Fund Convention (Article 5 (1)(2)). The decision to pay compensation through the COPE Fund — normally to every person having a valid claim under the international regime — was to be approved by the European Commission, hereby assisted by the COPE Fund Committee. However, the European Commission could decide not to pay compensation to the shipowner, manager or operator of the ship involved in the incident or to their representatives, or to any person in a contractual relationship with the carrier in respect of the carriage during which the incident occurred or any other person directly or indirectly involved in that carriage (Article 5 (4)). This exception had been provided to allow the European Commission a certain discretion as to the extent to which expenses by those most directly involved in the accident would be compensated. This mechanism aimed at ensuring the establishment of a relationship between the actual conduct of those involved and their right to compensation. In referring to the practice of the 1992 IOPC Fund — and very much unlike the original critics of the European Commission on the non-admissibility of claims for environmental damage under the international regime – the COPE Fund proposal did not foresee any compensation for environmental damage going beyond the 1992 Fund Convention[68].

[67] EUROPEAN COMMISSION, *Communication from the Commission to the European Parliament and the Council on a second set of community measures on maritime safety following the sinking of the oil tanker Erika*, 6 December 2000, COM (2000) 802def., 53-59.

[68] It can be noted that also Directive 2004/35/EC of the European Parliament and of the Council of 21 April 2004 on environmental liability with regard to the prevention and remedying of environmental

The aggregate amount of compensation payable by the COPE Fund would in respect of any one incident be limited, so that the total sum of that amount and the amount of compensation actually paid under the 1992 Civil Liability and Fund Convention for pollution damage within the scope of application of this regulation does not exceed 1,000 million euro (Article 5) This limit is similar to the limit of the *Oil Spill Liability Trust Fund* under OPA '90.
The COPE Fund – that would only be activated once an incident that has occurred in European waters exceeds, or threatens to exceed, the maximum limit provided by the IOPC Fund - would be financed by the European oil receivers: any person in a Member State who received more than 150,000 tonnes of contributing oil per year would have to pay its contribution to the COPE Fund, in a proportion which corresponds to the amount of oil received (Article 6).
The proposed regulation also included an article introducing financial penalties for grossly negligent behaviour by any person involved in the transport of oil by sea. This – non insurable - penalty would be imposed by Member States outside the scope of liability and compensation and would thus not be affected by any limitation of liability (Article 10).

3.3. Amendments

3.3.1. Amendments proposed by the Economic and Social Council (ESC) and the Committee of the Regions (COR)

The proposal was legally based on Article 80, para. 2 and Article 175, para. 1 EC Treaty and thus had to be adopted under the co-decision procedure. Both the ESC[69]and the COR[70] had been consulted. However, while they broadly approved the proposal in principle, both the ESC and the COR wanted to make the rules stricter.

damage (*OJ* L 143/56, 30 April 2004) does not apply to environmental damage or to any imminent threat of such damage arising from an incident in respect of which liability or compensation falls within the scope of any of the International Conventions listed in Annex IV, including any future amendments thereof, which is in force in the Member State concerned. It concerns the 1992 Civil Liability and Fund Conventions, the International Convention of 23 March 2001 on Civil Liability for Bunker Oil Pollution Damage (Bunker Oil Convention); the International Convention of 3 May 1996 on Liability and Compensation for Damage in Connection with the Carriage of Hazardous and Noxious Substances by Sea (HNS Convention); and the Convention of 10 October 1989 on Civil Liability for Damage Caused during Carriage of Dangerous Goods by Road, Rail and Inland Navigation Vessels (Article 4 (2) & Annex I Directive 2004/35/EC). See also *supra* in this book BRANS, E., *Estimating damages under the 2004 EC Directive on environmental liability* (Part I – Chapter 1).

[69] Opinion of the Economic and Social Committee of 30 May 2001 on the Proposal for a Directive of the European Parliament and of the Council establishing a Community monitoring, control and information system for maritime traffic, the Proposal for a Regulation of the European Parliament and of the Council on the establishment of a fund for the compensation of oil pollution damage in European waters and related measures, and the Proposal for a Regulation of the European Parliament and of the Council establishing a European Maritime Safety Agency, *OJ* C 221, 7 August 2001.

[70] Opinion of the Committee of the Regions of 13 June 2001 on the Proposal for a Regulation of the European parliament and of the Council on the establishment of a fund for the compensation of oil pollution damage in European waters and related measures, *OJ* C 357, 14 December 2001.

The ESC *inter alia* proposed that the compensation of damage caused to the environment should be reviewed and widened in the light of comparable compensation regimes established under Community law to cover claims concerning damages to biodiversity (para. 3.2.14.). Hereby too the ESC was highly critical that the damage eligible for compensation was the same as the one defined under the existing international conventions (para. 3.2.18.). Recalling the sinking of the chemical tanker *Ievoli Sun* the ESC highlighted the most unsatisfactory legal liability and compensation regime regarding hazardous and noxious substances other than oil (para. 3.2.16.). The introduction of penalties (legislation of a criminal nature) had been doubted to be compatible with the present stage of development of Community law. Therefore, the European Commission had been invited to produce an inventory of relevant national legislation in EU Member States before proceeding to the adoption of Article 10. Furthermore, the term "*grossly negligent acts or omissions*" had not been considered sufficiently precise for inclusion in a Community legal instrument: "*Pending developments regarding the Communitisation of the third pillar, it should be ensured that under national legislation of Member States there is no impunity for such offences*" (para. 4.3.1.2.).

Also the COR endorsed further provisions to be brought forward for pollution incidents other than oil (para. 4.2.). Hereby the COR stipulated that the proposals did not go "*to the real heart of trying to nail irresponsible ship owners who hide behind single ship companies, switch flags with apparent impunity, and abandon ships and their crews rather than meet their financial obligations when these vessels become embarrassing to them*" (para. 4.6.). Furthermore, the COR surged the European Commission and the Member States to establish better co-ordination of their activities in establishing a clear "chain of responsibility" so that liability for environmental clean up, or mitigating against further damage, could be secured efficiently.

It is worthwhile to mention that both the ESC and the COR urged the European Commission to develop the proposal for the establishment of a supplementary fund and higher compensation limits with the IMO.

3.3.2. Amendments proposed by the European Parliament

Pursuant to Article 251(2), Article 80(2) and Article 175(1) EC Treaty, the European Commission has submitted the initial COPE Fund Proposal to the European Parliament by letter of 8 December 2000. The proposal has then been referred to the Committee on Regional Policy, Transport and Tourism (RETT) and to the Committee on Budgets, the Committee on Industry, External Trade, Research and Energy and the Committee on the Environment, Public Health and Consumer Policy for their opinions. The RETT appointed Alain ESCLOPÉ as rapporteur at its meeting of 24 January 2001 and considered the Commission proposal and draft report at its meetings of 21 March, 25 April and 29 May 2001. At the last meeting it adopted the draft legislative resolution by 39 votes to 3, with 1 abstention. On 14

June 2001 the European Parliament endorsed the report and adopted a number of amendments to tighten the legislation[71].

The European Parliament, taking into account the work of the *Third Intersessional Working Group* under the auspices of the 1992 IOPC Fund (see below), endorsed the improvement of the protection of victims in case of a spill in Europe caused by oil, hazardous or other noxious substances by complementing the current *international* liability and compensation regime through the establishment of an international third tier[72].

The main amendments that had been adopted relate to the extension of the scope of application of the COPE Fund proposal to encompass other kinds of damage, *i.e.* bunker oil pollution damage and damage in connection with the carriage of hazardous and noxious substances by sea - whether carried by oil tankers or other ships[73]. These amendments had been justified by the non ratification of the 2001 Bunker Oil Convention and the 1996 HNS Convention and had major consequences for the text of the COPE Proposal. The COPE Proposal would provide for "*the Establishment of a Fund for Compensation for Tanker Pollution in European waters caused by oil, hazardous and noxious substances ("COPE Fund")*"[74]. Consequently, this COPE fund, which would be set up on top of the existing 1992 IOPC Fund, should thus consist of two layers, namely (i) "*Layer 1*" *consisting of higher limits of compensation to be paid by the shipowners where the cost of pollution damage exceeds or threatens to exceed the aggregate compensation limit under the existing 1992 Civil Liability and the 1992 Fund Conventions*; and (ii) "*Layer 2*" *consisting of a supplementary fund to be paid by the cargo receivers where the cost of pollution damage exceeds or threatens to exceed the aggregate compensation limit under the existing 1992 IOPC Fund as supplemented by the shipowner's contribution under Layer 1 of the COPE Fund or the 1996 HNS Convention once ratified.* Contributions should be borne by all operators involved in the transport of oil and of hazardous and noxious substances by sea instead of oil receivers alone. Furthermore, the European Parliament reduced the period for the collection of these contributions from one year to six months after the decision to levy contributions has been made by the European Commission[75].

---

[71] COMMITTEE ON REGIONAL POLICY, TRANSPORT AND TOURISM, *Report on the proposal for a European Parliament and Council regulation on the establishment of a fund for the compensation of oil pollution damage in European waters and related measures*, A5-0201/2001, 29 May 2001 and EUROPEAN PARLIAMENT, *Opinion of the European Parliament on the proposal for a European Parliament and Council regulation on the establishment of a fund for the compensation of oil pollution damage in European waters and related measures*, A5-0201/2001, 14 June 2001, *OJ* C 53E, 28 February 2002.

[72] Amendments n°s 5, 9 and 11.

[73] With the exception of nuclear materials.

[74] This title would have been confusing as the scope of application of both the 2001 Bunker Oil Convention and the 1996 HNS Convention is not restricted to tankers.

[75] Amendments n°s 1-8, 10, 12-18, 21-22, 24-28 and 44.

Moreover, the European Parliament included a new provision stating that "*the COPE Fund shall also pay compensation for damage caused to the environment when environmental costs are not covered by the international regime. These environmental costs are the costs for assessing the environmental damage of the incident and, if the environmental damage cannot be fully repaired, the costs of returning the environment to a state equivalent to that prior to the incident*"[76].

The European Parliament also introduced an amendment obliging the shipowner concerned with the tanker operation responsible for a marine pollution incident to contribute to the compensation of the victims on the same basis as the cargo receivers. To this end, each ship sailing in European territorial waters or marine economic interest zones should be able to prove it holds a financial guarantee or should pay a heavy financial penalty[77]. The European Parliament equally endorsed the development of clear procedural rules to ensure efficient interaction between the new COPE Fund and the existing international regime, so as to avoid imposing unnecessary administrative and cost burdens on either fund, and ensure a smooth transition from one fund to the other[78]. The European Parliament also proposed for the possibility of making an advance provisional payment within a period of six months[79]. Furthermore, chosen local representatives of the polluted area should take part in meetings of the COPE Fund Committee in an advisory capacity[80]. Finally, the COPE Fund Committee should submit an annual report on its activities to the Council of Ministers and the European Parliament in order to keep them appraised of its progress and activity [81].

The European Parliament rejected the provision empowering the European Commission to decide not to pay compensation to the shipowner, manager or operator of the ship involved in the incident or to their representatives or not to compensate any person in a contractual relationship with the carrier in respect of the carriage during which the incident occurred or any other person directly or indirectly involved in that carriage.

3.4. Amended COPE Fund Proposal

In an amended proposal[82] the European Commission agreed to a number of amendments that would clarify the text or the scope of the regulation or otherwise constitute editorial improvements. These amendments relate to the clarification of

---

[76] Amendment n° 51.
[77] Amendment n° 23.
[78] Amendment n° 32.
[79] Amendment n° 20.
[80] Amendments n°s 29 and 31.
[81] Amendments n°s 30 and 33.
[82] EUROPEAN COMMISSION, *Amended proposal for a Regulation of the European Parliament and of the Council on the establishment of a fund for the compensation of oil pollution damage in European waters and related measures*, 12 June 2002, COM (2002) 313 final.

the nature and activities of the COPE Fund, its link to the IOPC Fund, the facility to provide advance payment and the preparation of a progress report.

The European Commission also endorsed the provisional nature of the proposed regulation. In the preambule of the amended proposal the European Commission explicitly indicates that "*a considerably raise of the maximum amount of compensation available for oil spills could be done by complementing the existing international regime through the creation of an international supplementary fund. Until such an international fund is fully operational in all EU Member States concerned and provides adequate protection for accidents occurring in EU waters, a European Fund should be established to compensate claimants who have been unable to obtain full compensation under the international compensation regime, because the totality of valid claims exceed the amount of compensation available under the Fund Convention*".

Nevertheless, the European Commission rejected any amendment seeking to extend compensation for environmental damage. Moreover, the Commission has not accepted the amendments seeking to extend the scope of the COPE Fund to cover other forms of pollution than oil pollution by tankers and any amendment aiming at introducing an obligation for shipowners to pay a part of the compensation. The latter has been rejected on the basis that the existing international legal regime[83] does not allow additional compensation claims to be placed on the shipowner. Also the amendments proposing to increase the participation and role of the local representatives of the polluted region in the procedures of the COPE Fund Committee, which is a Management Committee under Decision 1999/468/EC, and the amendment proposing to limit the period of collection of the contributions to the COPE Fund from one year to six months had been rejected.

3.5. Latest evolutions in the COPE Fund Proposal

Although the majority of Member States agreed on the objective of creating a higher level of maximum cover through a top-up mechanism and the underlying principles of the proposed regulation, the general view among Member States was that it would be preferable to address this matter at an international level. The Transport Council of December 2000 adopted conclusions on the necessity to achieve ameliorations to the existing international regime, including "*a substantial increase in liability and compensation ceilings*". In June 2001 the Council adopted a common approach where Member States undertook to support and participate in a supplementary oil pollution compensation regime which should "*provide for full compensation*"[84].

---

[83] See Article III(4) 1992 Civil Liability Convention.

[84] EUROPEAN COMMISSION, *Proposal for a Council Decision authorising the Member States to sign, ratify or accede to, in the interest of the European Community, the Protocol of 2003 to the International Convention on the Establishment of an International Fund for Compensation for Oil Pollution Damage,*

Following the *Prestige* accident in November 2002, the Council took a stronger stance on the need for a top-up compensation arrangement. The European Summit on 21 March 2003 confirmed the views of various Council meetings in December in calling for: *"in terms of compensation for the victims of pollution, including environmental damage, Member States to pursue within the forthcoming diplomatic conference at the IMO in May an increase in the current ceiling on compensation to 1 billion euro; failing a positive outcome within the IMO to work on the existing proposal for a Regulation establishing a special European fund endowed with 1 billion euros with a view to the creation of the fund before the end of the year and drawing as much as possible on private funding"* As to the timing, paragraph 12 of the Transport Council Conclusions of 6 December 2002, subsequently endorsed by the Copenhagen Summit a week later, was even more clear: *REAFFIRMS the support of Member States to establish a supplementary compensation fund, developed in the IMO, to the benefit of the victims of oil pollution, which should be able to cover any future oil-spills up to EUR 1,000 million in the waters of the Member States of the EU foreseeing a rapid mechanism for payments and being operational by the end of 2003, and the intention of those Member States, which are parties to the existing global compensation regimes, to ratify the new supplementary fund. AGREES, in the event that the supplementary compensation fund is not established, to examine immediately a regulation on the establishment of a fund for the compensation of oil pollution damage in European waters, with the aim of establishing this fund before the end of 2003"*[85].

4. THE SUPPLEMENTARY FUND

4.1. Background: the work of the Third Intersessional Working Group

At its 4th extraordinary session (April 2000), the 1992 Fund Assembly has established a *Third Intersessional Working Group* (hereinafter called "Working Group") to assess the adequacy of the international compensation system created by the 1992 Civil Liability and Fund Conventions. The Working Group held its first session in July 2000, its second session in March 2001, its third session in June 2001, its fourth session in May 2002 and its fifth session in February 2003, all under the Chairmanship of Mr. Alfred POPP QC (Canada). Both 1971 Fund Member States, as well as States and Organizations which had observer status with the 1992 Fund Convention were invited to participate as observers[86]. Therefore, the European Commission has been granted observer status during the Working Group's sessions[87].

1992, and authorising Austria and Luxemburg, in the interest of the European Community, to accede to the underlying instruments, COM (2003) 534 final, 8 September 2003, 3-4.

[85] *Ibid.*

[86] 92FUND/A.7/4 – 92FUND/WGR.3/12.

[87] The 1992 Assemby has granted observer status to the European Economic Community at its 1st session in June 1996 (92FUND/A.1/34, para. 4.2). The European Commission had requested the 1992 Fund Assembly that the name to be used in referring to the observer in the IOPC Fund should be the European

In accordance with the mandate given by the Assembly, the Working Group held a general preliminary exchange of views concerning the need to improve the compensation regime provided by the 1992 Civil Liability and Fund Conventions, and drew up a list of issues which merit further considerations in order to ensure that the compensation systeem meets the needs of society.

4.1.1. Proposal for a two-layer third tier

During the preliminary exchange of views, several representatives endorsed the increase of the maximum limits for ensuring an adequate compensation to the victims of oil pollution at sea[88]. The Working Group stressed that these changes should be addressed in an international context. A number of delegations (Australia, Canada, Denmark, the Netherlands, Norway, Sweden and the United Kingdom) jointly forwarded their viewpoint on the occasion of the second session of the Working Group and introduced a twofold proposal. Firstly, the tacit acceptance procedure in the 1992 Civil Liability and Fund Conventions should be modified[89].

Commission. This would be more in line with with the situation in the IMO. The Assembly has decided at its 7th Session that the name to be used should be "European Commission" (92FUND/A.7/29, para. 31.2).

[88] 92FUND/WGR.3/2/1. Next to an increase of the maximum amounts under both the Civil Liability and the Fund conventions, various solutions have been suggested for ensuring an adequate compensation to the victims of oil pollution at sea: an increase in the limits available for compensation in the first and second categories (shipowners and insurers) (92FUND/WGR.3/2/2); a reduction of time-bar periods; a reduction in the duration of legal proceedings; the introduction of a guarantee in favour of the IOPC Fund to cover the risk of overpayments, the interruption of the time-bar periods by means of a simple formal request and the raising of the limits of the IOPC Fund by adding interest accrued on the amounts paid by the contributors (92FUND/WGR.3/5/3 & 92FUND/WGR.3/5/3/Corr. 1).

The Republic of Korea indicated that there would be considerable difficulties in setting up and maintaining a Fund of the type proposed by the EC without the involvement of the IOPC Fund. Therefore Korea submitted a proposal for the establishment of two new funds at a global level, namely a Cushion Fund and a Supplementary Fund, based on the assumption that the operation of the IOPC Fund currently in place should remain intact. The Cushion Fund and the Supplementary Fund would operate in addition to and in parallel with the current system based on the Civil Liability and Fund Conventions. Each Member State would have its own Cushion Fund which would be used only for the victims and contributors in that State. There would be no cross-subsidy and no mutuality. Funding would be based on the combination of a premium system where members are required to pay a fixed amount of premium calculated on the basis of potential claims and a call system (the current system for the amount of contributions). Three kinds of rates in the premium have been proposed. The proposal on the establishment of a Supplementary Fund as a third tier of compensation was fully based on the EC Proposal set in a global scenery (92FUND/WGR.3/5/4).

[89] It should be recalled that – simultaneously with the establishment of the Working Group – the tacit acceptance procedure, provided for in Article 15 of the 1992 Civil Liability Protocol and in Article 33 of the 1992 Fund Protocol, had been invoked, resulting in an increase of the compensation limits by 50 % in 2000. The delegations pointed out that "*the increase adopted by the Legal Committee in 2000 will take three years to come into force. Then, there will be a five year period before another increase can be considered. If a new increase is adopted at that point in time, it will again take three years for that increase to come into force. Hence, the total elapsed time between the increase adopted in 2000 and the next increase would be 11 years*. Two modifications have been proposed by the sponsors: *(i) to shorten the period from the adoption of the decision in the Legal Committee until the changes enter into force, so this period will be less than the existing three years, and (ii) to shorten or delete the time limit by which a further increase may be considered. These amendments would be included in a new Protocol to the 1992*

Secondly, the delegations proposed for the establishment of a voluntary "opt-in" third tier on top of the existing Civil Liability/Fund regimes, "*a supra tier*". The proposal was based on the idea that some Contracting States to the Civil Liability and Fund Conventions might wish to maintain the (increased) limits in the current 1992 regime, updated from time to time by the tacit amendment procedure, whilst other Contracting States might find it necessary to seek a solution which will ensure much higher limits in the foreseeable future and which are willing to impose higher costs on their industry. The initial proposal relates to the establishment of a new third tier of compensation that would be set up on top of the existing IOPC Fund 1992, consisting of *two* layers, namely "*layer 1*" that would establish higher limits of compensation to be paid by the shipowners, and "*layer 2*" that would establish a supplementary fund to be paid by the oil receivers[90]. The third tier could be adopted in a new – voluntary - protocol to both the Civil Liability and the Fund Conventions. The sponsors of the proposal highlighted that it would be important to bear in mind the balance between the obligations of the shipowners and the receivers of the contributing oil in setting up the third tier[91].

4.1.2. Critics

Although it was generally recognized that amendments to the provisions in the 1992 Civil Liability Convention relating to shipowners' liability would give rise to difficult treaty law issues, one of the critics made on the initial "*two-layer third tier*" proposal was that it would not be possible to include a layer of additional shipowner liability in the third tier without changing the 1992 Civil Liability Convention. Unless the 1992 Civil Liability Convention itself was changed, that Convention would prevent States Parties to the supplementary system from imposing the higher shipowner limits on ships flying the flag of States Parties to the 1992 Civil Liability

---

Conventions (92FUND/WGR.3/5/1). The 2000 amendments of 18 October 2000 came into force on 1 November 2003.

[90] Some options have been presented, seeking to meeting concerns of those who wish to maintain the present balance between shipping and oil interests in the international regime (92FUND/WGR.3/8/5). The first option relates to the introduction of a voluntary limit of liability for shipowners/insurers at the lower end of the scale of liability under the 1992 Civil Liability regime. This voluntary limit would apply only to those States parties to the third tier. Essentially this proposal would provide for an increase in the Civil Liability limit for incidents involving "small" tankers (92FUND/WGR.3/8/5, para. 3.1-3.3). A second option provides for a third tier split between receivers and shipowners, similar to the 1992 regime. For those States party to the third tier, this would increase the limits of liability for both Civil Liability and Fund regimes. Whilst the overall maximum amount would not differ, the initial burden of any limit set above the 1992 limit would be met by the shipowner, up to a limit. Above this limit the third tier fourth layer, as provided by the oil receivers; would meet the remaining compensation costs. This proposal would not work any differently to the present regime under the 1992 protocol (92FUND/WGR.3/8/5, para. 3.4). A third option relates to an alternative means of meeting the shipowners liability which would apply in tandem with the oil receivers' liability as part of the third tier above the 1992 Fund limit. Essentially, if any compensation costs were required above the 1992 Fund limit from States party to the third tier, these would – following a certain percentage - be met by both the shipowner and the receivers contribution at the same time, rather than an initial burden being placed, initially, on one single party (92FUND/WGR.3/8/5, para. 3.5-3.6). The fourth option relates to the "simple" increase of the liability limits under the 1992 Civil Liability and Fund Convention (92FUND/WGR.3/8/5, para. 3.7).

[91] 92FUND/WGR.3/5/1, para. 2.18-2.23.

Convention which did not become parties to the new third tier. The higher limits could only be imposed on ships flying the flag of a State Party to the new supplementary system or of a State not party to the 1992 Conventions. They feared that such a solution could result in shipowners choosing to flag out their ships from registers of State Parties to the new third tier to registers of States outside that system. For these reasons they considered that a third tier should be financed only by oil receivers[92].

4.1.3. Shipowners' liability

According to the OIL COMPANIES INTERNATIONAL MARINE FORUM (OCIMF) the introduction of an optional third tier wholly funded by the oil receivers would, in the event of a catastrophic event, cause an immediate and significant distorsion of the balance of risk between ship and cargo interests. Although OCIMF strongly believes that the optional third tier should consist of two parts so as to share the burden of contributions equitably between shipowner and cargo interests, OCIMF acknowledged that there might be scope for an interim solution with the introduction of an "opt-in" third tier (to be limited to a total of 400 million SDR), initially funded entirely by oil receivers in the opt-in countries and the coincident introduction by shipowners and their insurers of a voluntary scheme to increase the 1992 Civil Liability minimum liability treshold[93].

Representatives of shipowners and their insurers took the view that the issues relating to shipowners' liability should not be reopened since to do so would be detrimental to the position of victims of oil pollution[94]. Therefore, the INTERNATIONAL GROUP OF P & I CLUBS informed the Working Group of a study based on the Historical Tanker Spill Database 1990-1999 that concluded that the actual and inflated costs of all USA tanker and barge spills since the enactment of OPA'90 would have fallen within the existing 1992 IOPC Fund limit and that the present regime had resulted in an equitable sharing of burden between these two interests[95]. The INTERNATIONAL CHAMBER OF SHIPPING (ICS) pointed out that it should not be forgotten that the shipowner alone meets the great majority of claims without recourse to the IOPC Fund ad that ICS therefore does not feel that it

---

[92] 92FUND/WGR.3/8/4, para. 2.2.

[93] 92FUND/WGR.3/8/2, para. 2.4, 2.7 and 3.3.

[94] See also MAURA BARANDIARAN, J. (2003), *The International Regime on Liability and Compensation for Oil Pollution Damage: Recent Developments – Contribution to the European Workshop "Prestige: lessons learnt" (European Commission – DG Environment in collaboration with the Italian Presidency), held in Catania (Italy) on 13 November 2003*, http://europa.eu.int/comm/environment/civil/marin/prestige_catania/prestige_workshop_catania/session6/presentation_maura.pdf; INTERNATIONAL OIL POLLUTION COMPENSATION FUND (2002), *Report on the Activities of the International Oil Pollution Compensation Fund in 2001*, London, 32-34; INTERNATIONAL OIL POLLUTION COMPENSATION FUND (2003), *Report on the Activities of the International Oil Pollution Compensation Fund in 2002*, London, 31-32; INTERNATIONAL OIL POLLUTION COMPENSATION FUND (2004), *Report on the Activities of the International Oil Pollution Compensation Fund in 2003*, London, 33-34.

[95] 92FUND/WGR.3/8/3, para. 7, 13 and Annex.

is inequitable for oil receivers to contribute, substantially if necessary, in those numerically few cases where the shipowner's limits are exceeded[96]. The INTERNATIONAL GROUP OF P & I CLUBS introduced a proposal relating to a voluntary increase in the limit of liability for small ships under the 1992 Civil Liability Protocol which would apply only in those States which opt for the proposed third tier of compensation in excess of the 1992 IOPC Fund limits, funded by oil receivers[97]. Both the International Association of Independent Tanker Owners (INTERTANKO) and ICS have expressed the willingness to cooperate with the INTERNATIONAL GROUP OF P & I CLUBS in the development of its suggestion to voluntarily increase the limits of liability for smaller ships to around 20 million SDR in States which adopt the third tier, as a means of redressing any perceived imbalance between shipowners and oil receivers, should an opt-in third tier be established[98].

4.1.4. Draft Proposal

Based on these observations a draft proposal has been developed to supplement the 1992 Fund Convention with a voluntary protocol establishing a new compensation fund for oil pollution damage, the Supplementary Fund, to be financed by contributions from the oil receivers alone. The amount as to where the maximum limit should be set has been left blank in the Draft Protocol. The outcome of the Working Group's observations has been laid down in a revised text by the Director of the 1992 Fund and has been presented at the 6th Session of the 1992 Fund Assembly (October 2001). The revised text has then been discussed in depth resulting in some further amendments. It is worth mentioning that the Japanese delegation proposed the inclusion of a provision on capping of contributions (*cfr. infra*)[99]. The 1992 Fund Assembly has adopted the text of the draft protocol and instructed the Director of the 1992 IOPC Fund to submit the text of the draft protocol to the Secretary-General of the IMO requesting him to convene a Diplomatic Conference to consider the draft protocol at the earliest opportunity[100].
The questions relating to the shipowners' liability are still being discussed within the Working Group.

4.1.5. Future work

After having considered at its 6th session the Working Group's Report on its second and third sessions, the Assembly invited the Working Group to continue its exchange of views, including issues which have already been identified but not yet resolved, and to report to the next regular session of the Assembly. The main issues were as follows: (i) shipowners' liability; (ii) environmental damage; (iii) alternative dispute settlement procedures; (iv) non-submission of oil reports; (v) clarification of

---

[96] 92FUND/WGR.3/8/11.
[97] 92FUND/WGR.3/8/9.
[98] 92FUND/WGR.3/8/11 and 92FUND/WGR.3/8/16.
[99] 92FUND/A.6/4/6.
[100] 92FUND/A.6/28, para. 6.5-6.26 + Annex I.

the definition of 'ship'; (vi) application of the contribution system in respect of entities providing storage services; (vii) uniformity of application of the Conventions; and (viii) various issues of a treaty law nature[101].

Regarding "environmental damage" the Working Group agreed that an examination should be made of what could be achieved within the present definition of 'pollution damage' as regards the admissibility of claims for reinstatement of the environment and for the costs of environmental impact studies. It was also agreed that this should be considered in the context of a change to the IOPC Fund policy rather than as an amendment to the Conventions. A proposal to address the issues in an Assembly Resolution received considerable support in the Working Group. There was also support for an in-depth consideration of the issue of environmental damage in the longer term[102]. In April/May 2002 the Working Group considered the criteria to be applied with regard to the admissibility of claims for costs of post-spill environmental studies and for costs of measures of reinstatement of the polluted environment. The Working Group prepared a revised text of the relevant section of the Claims Manual, clarifying the criteria to be applied in respect of such claims, within the legal framework of the definition of 'pollution damage' in the 1992 Conventions. The revised text was approved by the Assembly at its October 2002 session[103]. The revised version of the Claims Manual was published in November 2002[104].

4.2. The International Conference on the Establishment of a Supplementary Fund for Compensation for Oil Pollution Damage

In accordance with Article 2(b) of the Convention on the International Maritime Organisation, the Council of the IMO, at its twenty-first extraordinary session (16 November 2001), authorized the convening of a diplomatic conference of Contracting States to the International Convention on the Establishment of an International Fund for Compensation for Oil Pollution Damage, 1992, to consider the adoption of a draft protocol to the 1992 Fund Convention. The IMO Assembly, at its twenty-second regular session, endorsed this decision by resolution A.906(22)

---

[101] 92FUND/A.7/4.

[102] INTERNATIONAL OIL POLLUTION COMPENSATION FUND (2002), *o.c.*, 34.

[103] 92FUND/A.7/29.

[104] INTERNATIONAL OIL POLLUTION COMPENSATION FUND (2003), *o.c.*, 30. See also *supra* in this book, NICHOLS, J., *Scope of compensation for environmental damage under the 1992 Civil Liability Convention and the 1992 Fund Convention* (Part II – Chapter 1) and GONSAELES, G. & MAES, F. (2002), Taak V. Het ontwikkelen van technische en juridische procedures met het oog op de evaluatie van degradatie aan het mariene milieu en het financieel verhaal op de vervuiler: Het internationaal aansprakelijkheidsregime voor schade bij olieverontreiniging op zee: onderzoek naar de vergoedbaarheid van milieuschade, in: MAES, F. (Ed.), *MARE-DASM "Duurzaam Beheer van de Noordzee" – Plan voor wetenschappelijk ondersteuning van een beleid gericht op duurzame ontwikkeling, studie in opdracht van de Diensten van de Eerste Minister, Federale Diensten voor Wetenschappelijke, Technische en Culturele Aangelegenheden*, Ghent, 143 p.

of 29 November 2001 on the Work Programme and Budget for the Twenty-Second Financial Period 2002-2003[105].

The basic proposal for discussion by the Conference consists of the draft protocol to the International Convention on the Establishment of an International Fund for Compensation for Oil Pollution Damage, 1992, prepared by the 1992 IOPC Fund[106]. The IMO Legal Committee, at its eighty-fourth session (22-26 April 2002), approved the draft protocol and decided that it was ready for submission to a diplomatic conference and that it had good prospects both for adoption by a conference and subsequent implementation by States.

The International Conference on the Establishment of a Supplementary Fund for Oil Pollution Damage has been convened in London under the auspices of the IMO from 12 to 16 May 2003. The Conference was attended by representatives of 51 Contracting States to the 1992 Fund Convention and by observers of 17 IMO Member States and 1 IMO Associate Member (Hong Kong Special Administrative Region), whereof 3 State Parties to both the 1992 Civil Liability and Fund Conventions. Two inter-governmental organizations which have concluded agreements of cooperation with the IMO — the IOPC Fund and the European Commission — and 12 non-governmental organizations which have been granted consultative status with the IMO also attended the Conference as observer[107].

As a result of its deliberations, on 16 May 2003 the International Conference adopted the *Protocol of 2003 to the International Convention on the Establishment of an International Fund for Compensation for Oil Pollution Damage, 1992* (hereinafter called "the 2003 Protocol" or "the Supplementary Fund Protocol")[108]. The International Conference also adopted three resolutions[109]: the *Resolution on Financing of the International Conference to adopt a draft protocol to the International Convention on the Establishment of an International Fund for Compensation for Oil Pollution Damage, 1992;* the *Resolution on Establishment of the International Oil Pollution Compensation Supplementary Fund;* and the *Resolution on Review of the International Compensation Regime for Oil Pollution Damage for Possible Improvement.*

---

[105] *Id est "provided that the Conference should not result in any costs and expenses to IMO"*. See also Resolution on Financing of the International Conference to adopt a draft protocol to the International Convention on the Establishment of an International Fund for Compensation for Oil Pollution Damage, 1992 (LEG/CONF.14/22).

[106] LEG/CONF.14/3.

[107] Final Act of the International Conference on the Establishment of a Supplementary Fund for Compensation for Oil Pollution Damage (LEG/CONF.14/21). The list of participants of the IMO Secretariat also included the Democratic People's Republic of Korea as an observer (LEG/CONF.14/INF.1).

[108] LEG/CONF.14/20.

[109] LEG/CONF.14/22.

4.3. The Protocol of 2003 to the International Convention on the Establishment of an International Fund for Compensation for Oil Pollution Damage, 1992

4.3.1. Objectives

Within the preamble of the Protocol reference is made to the willingness of a number of Contracting States to the 1992 Civil Liability and Fund Conventions to urgently make available additional funds for compensation through the creation of a supplementary scheme to which States may accede if they so wish. Hereby the voluntary character of the Protocol has been stressed. The supplementary scheme should seek to ensure that victims of oil pollution damage are compensated in full for their loss or damage and should also alleviate the difficulties faced by victims in cases where there is a risk that the amount of compensation available under the 1992 Liability and 1992 Fund Conventions will be insufficient to pay established claims in full and that as a consequence the 1992 IOPC Fund has decided provisionally that it will pay only a proportion of any established claim[110].

For achieving these objectives an International Supplementary Fund for compensation for pollution damage, to be named "*The International Oil Pollution Compensation Supplementary Fund, 2003*" has been established[111]. The Supplementary Fund will have an Assembly and a Secretariat headed by a Director, that will work according to the rules of the 1992 IOPC Fund. The Secretariat and the Director of the 1992 IOPC Fund may also perform the function of Secretariat and the Director of the Supplementary Fund[112]. This Supplementary Fund has to be recognized as a legal person, legally represented by the Director of the Supplementary Fund[113].

4.3.2. Scope of application of the 2003 Protocol

The narrow link between the 1992 Fund Convention and the 2003 Supplementary Fund is also reflected within the scope of application of the 2003 Protocol. Identical to the 1992 Civil Liability and Fund Conventions, the protocol applies exclusively to pollution damage caused: (i) in the territory, including the territorial sea, of a Contracting State, and (ii) in the exclusive economic zone of a Contracting State, established in accordance with international law, or, if a Contracting State has not

---

[110] Article 4(5) of the 1992 Fund Convention provides for a proportionate distribution ("*pro rata*") payments of the available amounts when the amount of established claims against the 1992 IOPC Fund exceeds the aggregate amount of compensation.

[111] Article 2(1) Supplementary Fund Protocol 2003. The preparations for the establishment of the Supplementary Fund are to be made by the Director of the 1992 IOPC Fund upon instructions of the 1992 IOPC Fund Assembly (Resolution on Establishment of the International Oil Pollution Compensation Supplementary Fund as contained in LEG/CONF.14/22). See also 92FUND/A.8/4/1.

[112] Articles 16 & 17 Supplementary Fund Protocol 2003. The Protocol also provides for provisions relating to possible conflicts of interests between the 1992 IOPC Fund and the Supplementary Fund. See also the Resolution on Establishment of the International Oil Pollution Compensation Supplementary Fund (LEG/CONF.14/22).

[113] Article 2(2) Supplementary Fund Protocol 2003.

established such a zone, in an area beyond and adjacent to the territorial sea of that State determined by that State in accordance with international law and extending not more than 200 nautical miles from the baselines from which the breadth of its territorial sea is measured; and to preventive measures, wherever taken, to prevent or minimise such damage[114]. All definitions on "*ship*", "*person*", "*owner*", "*oil*", "*pollution damage*", "*preventive measures*" and "*incident*" have the same meaning as in article I of the 1992 Civil Liability Convention and Article 1 of the 1992 Fund Convention[115]. This implies that the 2003 Protocol does not change the meaning of the notion "*pollution damage*", so that compensation for environmental damage is to be restricted to the costs of reasonable measures of reinstatement actually undertaken or to be undertaken.

4.3.3. Amount of compensation

Compensation limit. The aggregate amount of compensation payable by the Supplementary Fund in respect of any one incident is limited to 750 million units of account (to be converted into national currency on the basis of the value of that currency by reference to the Special Drawing Right (SDR) on the date determined by the 1992 Fund Assembly for conversion of the maximum amount payable under the 1992 Civil Liability and Fund Conventions)[116]. In calculating the compensation limit of 750 millions SDR payable by the Supplementary Fund in respect of a given incident, the compensation actually paid under the 1992 Civil Liability Convention and the 1992 Fund Convention must be taken into account[117].

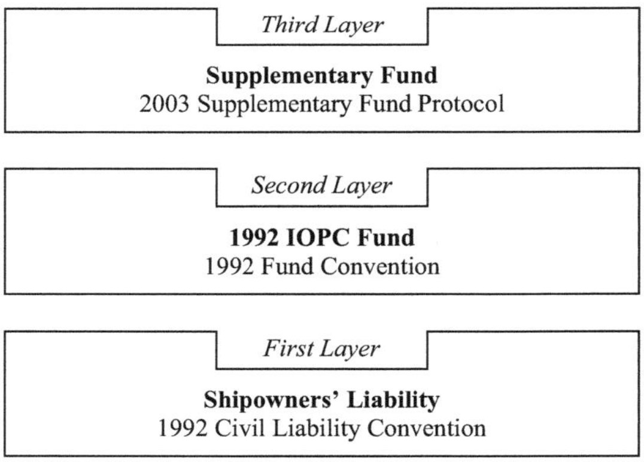

[114] Article 3 Supplementary Fund Protocol 2003.
[115] Article 1(6) Supplementary Fund Protocol 2003.
[116] Article 4(2) Supplementary Fund Protocol 2003. This compensation limit equals € 889,042,500 or £ 615,243,000 or $ 1,103,565 as converted by the rates of 1 October 2004 as defined by the International Monetary Fund. At the day of adoption of the 2003 Supplementary Fund Protocol the 750 million SDR compensation limit equalled € 916,515,000 or £ 649,545,000 or $ 1,053,262.
[117] Article 4(2)a Supplementary Fund Protocol 2003.

Amendments. The 2003 Protocol has anticipated calls for higher compensation limits by the introduction of a tacit acceptance procedure for amendments to the compensation limit[118]. The tacit acceptance procedure is very similar to the procedure set out in Article 15 of the 1992 Civil Liability Protocol and Article 33 of the 1992 Fund Protocol. A proposal to amend the limits of the amounts of compensation must be made upon the request of at least one quarter of the Contracting States and must be circulated to all IMO Members and to all Contracting States[119]. If the proposal is adopted in the Legal Committee of IMO by a two-thirds majority of the Contracting States (whether IMO Members or not) present and voting, on condition that at least one half of the Contracting States are be present at the time of voting, the amendment can enter into force 24 months after its adoption[120]: the amendment shall be deemed to have been accepted 12 months after the date of notification, unless within that period at least one quarter of Contracting States at the time of the adoption of the amendment by the Legal Committee have communicated to the IMO that they do not accept the amendment, in which case the amendment is rejected and shall have no effect[121]. An amendment deemed to have been accepted shall enter into force 12 months after its acceptance[122]. All Contracting States shall be bound by the amendment, unless they denounce the 2003 Protocol at least 6 months before the amendment enters into force. Such denunciation shall take effect when the amendment enters into force[123].

In amending the compensation limit, the Legal Committee must take into account the experience of incidents and in particular the amount of damage resulting there from and changes in monetary values[124]. The determination of a higher compensation limit through the tacit acceptance procedure is subject to certain conditions. Two options are possible, depending on if the compensation limits laid down in the 1992 Fund Convention have been increased.

If the compensation limit laid down in the 1992 Fund Convention has been increased by a protocol hereto, the compensation limit of the Supplementary Fund may be increased by the same amount through the tacit acceptance procedure without further restrictions[125]. The Supplementary Fund Protocol explicitly refers to an increase of the limits laid down in the 1992 Fund Convention *by a protocol*. This seems to exclude an increase of the limits according to the tacit acceptance procedure as laid down in Article 33 of the 1992 Fund Protocol as these amendments do not take the form of a protocol but of a Resolution of the Legal Committee of the IMO.

---

118 Articles 24-25 Supplementary Fund Protocol 2003.
119 Article 24(1)(2)(3) Supplementary Fund Protocol 2003.
120 Article 24(4) Supplementary Fund Protocol 2003.
121 Article 24(7) Supplementary Fund Protocol 2003.
122 Article 24(8) Supplementary Fund Protocol 2003.
123 Article 24(9) Supplementary Fund Protocol 2003.
124 Article 24(5) Supplementary Fund Protocol 2003.
125 Article 25(1) Supplementary Fund Protocol 2003.

If the proposed amendments do not result from an increase of the 1992 Fund compensation limit by a protocol hereto, the increase of the compensation limit as laid down in the 2003 Protocol is subject to three cumulative conditions: (i) no amendments to the compensation limit may be considered before the date of entry into force of the 2003 Protocol nor less than three years from the date of entry into force of a previous amendment[126]; (ii) the proposed compensation limit may not be increased so as to exceed an amount which corresponds to the limit laid down in the 2003 Protocol — *i.e.* 750 million SDR — increased by 6 %/year calculated on a compound basis from the date when this Protocol is opened for signature to the date on which the Legal Committee's decision comes into force; and (iii) the compensation limit may not be increased so as to exceed an amount which corresponds to the limit laid down in the 2003 Protocol multiplied by three (*i.e.* 2.250 SDR)[127]. However, if the limit as laid down in the 2003 Protocol has formerly been increased as a consequence of an increase of the compensation limit laid down in the 1992 Fund Convention (as described above), any subsequent amendment shall be calculated on the basis of the new limit[128].

4.3.4. Supplementary Compensation

The Supplementary Fund shall pay compensation to any person suffering pollution damage if such person has been unable to obtain full and adequate compensation for an established claim for such damage under the terms of the 1992 Fund Convention[129]. For the purposes of the 2003 Protocol, an "*established claim*" means a claim (i) which has been recognized by the 1992 IOPC Fund or been accepted as admissible by decision of a competent court binding upon the 1992 IOPC Fund not subject to ordinary forms or review, *and* (ii) which would have been fully compensated if the 1992 IOPC Fund compensation limit had not been applied to that incident[130]. With this provision the practice of the 1992 IOPC Fund towards compensation has been transfered to the 2003 Protocol.

The Supplementary Fund regime will only be activated if the 1992 Fund Assembly has considered that the total amount of established claims exceeds, or there is a risk that the total amount of the established claims will exceed, the aggregate amount of compensation available under the 1992 Fund Convention in respect of any one incident[131] and that as a consequence the 1992 Fund Assembly has decided

---

[126] Article 24(6)a Supplementary Fund Protocol 2003. However, nothing prevents the Contracting States to amend the limit as laid down in the 2003 protocol by amending the Protocol itself (see also Articles 39, 40 Vienna Convention on the Law of the Treaties).

[127] Article 24(6)b and (6)c Supplementary Fund Protocol 2003.

[128] Article 25(2) Supplementar Fund Protocol 2003.

[129] Article 4(1) Supplementary Fund Protocol 2003.

[130] Article 1(8) Supplementary Fund Protocol 2003.

[131] The applicable limit of compensation, laid down in article 4 (4) of the 1992 Fund Convention – including the increase of the compensation limits by 50 % in 2000 under the tacit acceptance procedure - currently amounts to 203 million SDR for incidents occurring from 1 November 2003 under the 1992 regime (€ 240.6 million or £ 166.5 million or $ 298.7 million as converted by the rates of 1 October 2004 as defined by the International Monetary Fund).

provisionally or finally that payments will only be made for a proportion of any established claim. The Supplementary Fund Assembly shall then decide whether and to what extent the Supplementary Fund shall pay the proportion of any established claim not paid under the 1992 Civil Liability and Fund Conventions[132].

Although the compensation limit laid down in the Supplementary Fund Protocol is considered worldwide as sufficient to cover any compensation as a result of an oil spill, the 2003 Protocol has foreseen in a *pro rata* distribution of the available amount if the amount of established claims exceeds the aggregate amount of compensation payable in respect of any one incident (*i.e.* 750 million SDR): "*the amount available shall be distributed in such a manner that the proportion between any established claim and the amount of compensation actually recovered by the claimant under this Protocol shall be the same for all claimants*"[133].

The legal provisions governing actions for compensation, including the (exclusive) jurisdictional competence of courts under the Supplementary Fund regime, have been integrated as much as possible within the 1992 Civil Liability and Fund regime[134]. A claim for compensation made against the 1992 IOPC Fund is to be regarded as a claim made by the same claimant against the Supplementary Fund[135], thus avoiding to put an extra administrative burden on the claimant. Rights to compensation against the Supplementary Fund shall be extinguished only if they are extinguished against the 1992 IOPC Fund under the 1992 Fund Convention[136]: claimants ultimately lose their right to compensation under the 1992 Fund Convention unless they bring court action against the 1992 Fund within three years of the date on which the damage occurred, or make formal notification to the 1992 IOPC Fund of a court action against the shipowner or his insurer within that three-year period. Although damage may occur some time after an incident takes place, court action must in any case be brought within six years of the date of the incident[137].

Actions for compensation against the Supplementary Fund should be brought before a court competent under article IX of the 1992 Civil Liability Convention against the owner of a ship or his guarantor. Such court shall have exclusive jurisdictional competence over any action for compensation against the Supplementary Fund. However, where an action for compensation for pollution damage under the 1992 Civil Liability Convention has been brought before a court in a Contracting State to the 1992 Civil Liability Convention or to both the 1992 Civil Liability and Fund

---

[132] Article 5 Supplementary Fund Protocol 2003.
[133] Article 4(3) Supplementary Fund Protocol 2003.
[134] Article 6 & 7(1) Supplementary Fund Protocol 2003.
[135] Article 6(2) Supplementary Fund Protocol 2003.
[136] Article 6(1) Supplementary Fund Protocol 2003.
[137] INTERNATIONAL OIL POLLUTION COMPENSATION FUND (2002), *International Oil Pollution Compensation Fund 1992 - Claims Manual*, London, 13.

Conventions but not to the 2003 Protocol, any action against the Supplementary Fund should be brought either before a court of the State where the Supplementary Fund has its headquarters or before any court of a Contracting State to this Protocol competent under article IX of the 1992 Civil Liability Convention[138].

Any judgment given against the Supplementary Fund by such court shall, when it has become enforceable in the State of origin and is in that State no longer subject to ordinary forms of review, be recognized and enforceable in each Contracting State on the same conditions as are prescribed in article X of the 1992 Civil Liability Convention[139], *id est* except where the judgment was obtained by fraud or where the defendant was not given reasonable notice and a fair opportunity to present his case. A judgment shall be enforceable in each Contracting State as soon as the formalities required in the State have been complied with. The formalities shall not permit the merits of the case to be re-opened. Article 8(2) of the 2003 Protocol contains a provision allowing a Contracting State to apply other rules for the recognition and enforcement of judgments, provided that their effect is to ensure that judgments are recognized and enforced at least to the same extent as described above. This provision has been introduced into the 2003 Protocol upon request of the European Commission because Articles 7 and 8 relate to the jurisdiction, recognition and enforcement of judgments[140]. These provisions touch upon the provisions of Regulation 44/2001 on jurisdiction and the recognition and enforcement of judgments in civil and commercial matters, as amended[141], being an exclusive competence of the EU since the entry into force of the Treaty of Amsterdam. All Member States, other than Denmark[142], are bound by the provisions of Regulation 44/2001. In the view of the Commission, in accordance with the case law of the Court of Justice, Member States, whether acting individually or collectively, lose their right to assume obligations with third countries as and when common rules which could be affected by those obligations come into being. It follows that only the European Community is competent for the negotiation, conclusion and fulfillment of such international commitments. The articles dealing with the jurisdiction, recognition and enforcement of judgments should be compatible, to the

[138] Article 7(2)(3) Supplementary Fund Protocol 2003.

[139] Article 8(1) Supplementary Fund Protocol 2003.

[140] LEG/CONF.14/7. Similar provisions on the jurisdiction, recognition and enforcement of judgments within the Bunker Oil Convention 2001 and the HNS Convention 1996 that had not been brought to the attention of the European Commission in due time have urged the European to provide for a decision authorizing the Member States to sign and ratify the Conventions (cfr. Decision 2002/762/EC of 19 September 2002 authorizing the Member States, in the interest of the Community, to sign, ratify or accede to the International Convention on Civil Liability for Bunker Oil Pollution Damage, 2001 (*OJ* L 256, 25 September 2002) and Council Decision 2002/971/EC of 18 November 2002 authorising the Member States, in the interest of the Community, to ratify or accede to the International Convention on Liability and Compensation for Damage in Connection with the Carriage of Hazardous and Noxious Substances by Sea, 1996 (*OJ* L 337, 13 December 2002).

[141] Regulation 44/2001/EC of 22 December 2000 on the jurisdiction and the recognition and enforcement of judgments in civil and commercial matters, *OJ* L 12, 16 January 2001.

[142] The Brussels Convention of 27 September 1968 on jurisdiction and the enforcement of judgments in civil and commercial matters (*OJ* C 189, 28 July 1990) will continue to apply to Member States bound by the Convention that are excluded from the Regulation.

extent possible, with Community legislation in this area. Although Article 7 of the Supplementary Fund Protocol - in contrast to the multiple grounds of jurisdiction available under Regulation 44/2001 - mandates the exclusive jurisdiction of the State Party where pollution damage occurred as a main rule, the European Commission has pointed out the very specific nature of the Supplementary Protocol. The variation between the oil pollution system and the Community regime as regards the competent jurisdiction was thus not be addressed in the context of the Supplementary Fund Protocol. As to the rules on the recognition and enforcement of judgments, the Commission considered that a continued application of Chapter III of Regulation 44/2001 between Member States is essential, when it comes to the recognition and enforcement of judgments given by a court of a Member State in another Member State[143].

Provisions on subrogation are identical to the ones laid down in Article 9 of the 1992 Fund Convention. However, the Supplementary Fund, in respect of any amount of compensation for pollution damage paid by the Supplementary Fund, not only acquires by subrogation the rights that the person so compensated may enjoy under the 1992 Civil Liability Convention against the owner or his guarantor, but also the rights that the person compensated by it may enjoy under the 1992 Fund Convention against the 1992 IOPC Fund[144].

4.3.5. Funding of the Supplementary Fund

Similarly to the 1971 and 1992 Fund Conventions, the Supplementary Fund is to be financed by means of annual contributions to be made in respect of each Contracting State by any person who has received total quantities of contributing oil exceeding 150,000 tons (i) that has been carried by sea to the ports or terminal installations in the territory of a Contracting State; or (ii) that has been received in any installations situated in the territory of a Contracting State after it has been carried by sea and discharged in a port or terminal installation of a non-Contracting State. For the latter only the *first* receipt of contributing oil in a Contracting State after its discharge in a non-Contracting State will be taken into account[145]. However, for the purpose of the 2003 Protocol, there shall be deemed to be a minimum receipt of 1 million tons of contributing oil in each Contracting State. When the aggregate quantity of contributing oil in a Contracting State is less than 1,000,000 tons, the Contracting State will assume the obligations that would be incumbent on any person who would be liable to contribute to the Supplementary Fund in so far as no liable person exists

[143] Proposal for a Council Decision authorizing the Member States to sign, ratify or accede to, in the interest of the European Community, the Protocol of 2003 to the International Convention on the Establishment of an International Fund for Compensation for Oil Pollution Damage, 1992, and authorizing Austria and Luxemburg, in the interest of the European Community, to accede to the underlying instruments, COM (2003) 534 final, 8 September 2003, 5-6.

[144] Article 9 Supplementary Fund Protocol 2003.

[145] Article 10(1) Supplementary Fund Protocol 2003.

for the aggregated quantity of oil received[146]. After the Supplementary Fund Assemble has decided the total amount of contributions to be levied, the Director of the Supplementary Fund will calculate for each such person the amount of that person's annual contribution[147].

The 2003 Supplementary Fund Protocol provides for an interesting provision relating to "*capping*" as a transitional measure. This amendment has been introduced into the protocol on the demand of Japan to avoid an excessive burden for a few specific Contracting States and to secure a fair burden among them[148]. The aggregate amount of the annual contributions payable in respect of contributing oil received in a single Contracting State during a calendar year shall not exceed 20 % of the total amount of annual contributions pursuant to the 2003 Protocol in respect of that calendar year. If the aggregate amount of the contributions payable by contributors in a single Contracting State in respect of a given calendar year exceeds 20 % of the total annual contributions, the contributions payable by all contributors in that State shall be reduced *pro rata* so that their aggregate contributions equal 20 % of the total annual contributions to the Supplementary Fund in respect of that year. In that case the contributions payable by persons in all other Contracting States shall be increased *pro rata* so as to ensure that the total amount of contributions payable by all persons liable to contribute to the Supplementary Fund in respect of the calendar year in question will reach the total amount of contributions decided by the Assembly. This transitional measure shall operate until the total quantity of contributing oil received in all Contracting States in a calendar year - including the assumption that there shall be a minimum receipt of 1 million tons of contributing oil in each Contracting State - has reached 1,000 million tons or until a period of 10 years after the date of entry into force of this Protocol has elapsed, whichever occurs earlier[149].

---

[146] Article 14 Supplementary Fund Protocol 2003.

[147] Article 11 Supplementary Fund Protocol 2003.

[148] LEG/CONF.14/9. It is to be recalled that Japan, receiving 18.51 % of contributing oil (242.801.951 tonnes) in the calendar year 2003, is the main contributor to the 1992 IOPC Fund (INTERNATIONAL OIL POLLUTION COMPENSATION FUND (2004), *o.c.*, 155). A proposal from the Republic of Korea – being the third contributor to the 1992 IOPC Fund with 114,661,792 tons of contributing oil received in 2003 – to change the rule to apply a single and same rate for all oils in calculating contributions, and to introduce for the purpose of the Supplementary Fund, different rates in calculating contributions, in reflection of the geographical location and economic development of each State, has not been accepted. Korea considered that in a nation of a developing region, it would be rare or extremely rare to find an incident for which the total amount of claims amounts to or even comes close to the limit prescribed under the 1992 Fund Convention. The amounts of actual loss suffered are bound to vary depending on the amount of income and the costs to restore. This means that the Fund is required to pay a much higher amount as compensation for a similar incident in a nation in a well-developed region than for a nation in a developing region and leads to a manifestly unjust result from the viewpoint of the developing nations, since a contributor in a developing nation pays contributions at the same rate as a contributor in a developed nation with an equal limitation on the amount of compensations, while victims in a developing nation are bound to receive less amount of compensation than those in a developed nation due to the differences in the level of income and costs to restore (LEG/CONF.14/10).

[149] Article 18 Supplementary Fund Protocol 2003.

For the operationalization of these provisions, it is necessary that every Contracting State will communicate to the Director of the Supplementary Fund information on oil receipts (provided that communications made to the Director of the 1992 IOPC Fund shall be deemed to have been made also to the Director of the Supplementary Fund)[150]. The Contracting State will equally inform the Director if there is no person having received a total quantity of contributing oil exceeding 150,000 tons[151].

The non-submission of the "oil reports" by State Parties to the IOPC Fund has appeared to be problematic[152]. Therefore, the 1992 Fund Convention already provided that Contracting States will be liable to compensate the IOPC Fund for financial losses, if these losses are the result of the non-fulfilment of the obligation to submit the information on oil receipts to the Director. Furthermore, the 1992 Fund Convention would not come into force for a Contracting State until the names and addresses of any person who in respect of that State would be liable to the Fund as well as data on the relevant quantities of contributing oil received by any such person in the territory of that State during the preceding calendar year have been submitted at the moment of adoption and annually thereafter. The 2003 Protocol not only repeats these provisions[153], but also provides for stronger measures, stipulating that no compensation shall be paid by the Supplementary Fund for pollution damage in respect of a given incident or for preventive measures to prevent or minimize such damage in the territory of a Contracting State, until the obligations to communicate the "*oil reports*" to the Director have been complied with in respect of that Contracting State for all years prior to the occurrence of that incident. The circumstances under which a Contracting State shall be considered as having failed to comply with its obligations will be determined by the Supplementary Fund Assembly[154]. Where compensation has been denied *temporarily*, such compensation will be denied *permanently* in respect of that incident if these treaty obligations have not been complied with one year after the Director has notified the State of its failure to report[155]. This extension within the 2003 Protocol will probably be an effective means of enforcement for the information on contributing oil.

4.3.6. Entry into force and future of the Supplementary Fund Protocol, 2003

To stress the voluntary character of the Supplementary Fund, the International Conference has opted for a separate protocol of the 1992 Conventions, to be ratified,

[150] Article 13(1) Supplementary Fund Protocol 2003.
[151] Article 15(1) Supplementary Fund Protocol 2003.
[152] For instance no report has been submitted in the calendar year 2003 from Angola, Bahamas, Bahrain, Cambodia, Colombia, Comoros, Congo, Djibouti, Dominica, Dominican Republic, Gabon, Georgia, Guinea, Kenya, Madagascar, Morocco, Namibia, Nigeria, Papua New Guinea, Samoa, Sierra Leone, Saint Vincent and the Grenadines and Tanzania (INTERNATIONAL OIL POLLUTION COMPENSATION FUND (2004), *o.c.*, 155).
[153] Article 15(4) 1992 Fund Convention & Article 29 1992 Fund Protocol; Article 13 (2) & 20 Supplementary Fund Protocol 2003.
[154] Article 15(2) Supplementary Fund Protocol 2003.
[155] Article 15(3) Supplementary Fund Protocol 2003.

accepted, approved or acceded to *only* by States that are a party to the 1992 Fund Convention[156]. The entry into force of the 2003 Protocol depends on two cumulative conditions. Firstly, at least 8 States that have signed the protocol without reservation, must have deposited instruments of ratification, acceptance, approval or accession with the Secretary-General of the IMO. Secondly, the total quantity of contributing oil must be at least 450 million tons, including the assumption that there shall be a minimum receipt of 1 million tons of contributing oil in each Contracting State. The 2003 Protocol will enter into force 3 months following the date of the deposit by such State of the appropriate instrument[157].

The Supplementary Fund Protocol entered into force on 3 March 2005. To date, eight 1992 Fund Member States (Denmark, France, Norway, Finland, Ireland, Spain, Germany and Japan) ratified the 2003 Supplementary Fund Protocol[158]. The total amount of contributing oil received in these States during 2002, the most recent year for which figures are currently available, was 479, 195, 151 tons, with a minimum receipt of 4, 432, 177 tons (Ireland)[159]. Portugal will join by 15 May 2005, thus increasing the total amount of contributing oil to 493, 699, 271 tons.

The considerations made on the occasion of the work of the Third Intersessional Working Group and at the International Conference clearly show that it is of major importance to maintain an equitable sharing of the financial burden of oil spills between cargo receivers and shipowners. Therefore, the Supplementary Fund Protocol is only to be seen as an interim solution in the light of a further review of the international compensation regime for oil pollution damage, especially regarding the relationship between shipower liability and oil receiver contributions. Moreover, the International Conference also adopted a Resolution on the review of the international compensation regime for oil pollution damage for possible improvement[160].

Bearing in mind this equitable sharing of financial burden, the Club Boards of the P & I Clubs approved the main principles of a proposed agreement amongst shipowners (known as *Small Tanker Owners' Pollution Indemnification Agreement* - STOPIA), to voluntarily increase the minimum limit of liability for small ships. The principal elements of the STOPIA scheme which have been agreed by the Club Boards are as follows:

(i) under STOPIA the owners of relevant tankers of 29,548 GT or less would contract with the IOPC Fund to reimburse claims paid in excess of the relevant limit of liability under 1992 Civil Liability Convention up to 20 million SDR per incident;

---

156 Article 19 Supplementary Fund Protocol 2003.
157 Article 21 Supplementary Fund Protocol 2003.
158 For an updated state of ratifications, please consult http://www.imo.org, http://www.iopcfund.org or *Ratification of Maritime Conventions*.
159 INTERNATIONAL OIL POLLUTION COMPENSATION FUND (2004), *o.c.*, 155.
160 LEG/CONF.14/22.

(ii) the scheme would apply to approximately 6,000 tank vessels, representing about 75 % of the world fleet of tankers falling within the definition of "ship" within the 1992 Civil Liability Convention;

(iii) the scheme would only apply in the event of a tanker spill affecting a State party to the Supplementary Fund when liability was imposed under 1992 Civil Liability Convention;

(iv) the scheme would come into effect at the same time as the entry into force of the Supplementary Fund;

(v) the flag of the vessel or the ownership of the cargo would not be relevant;

(vi) the compensation limit as laid down in the 1992 Civil Liability Convention would have to be exceeded, but the scheme would operate even if claims did not reach the third tier Supplementary Fund;

(vii) the tanker owner's liability under the scheme would not exceed the 1992 Civil Liability limit plus the voluntary tranche;

(viii) the Clubs would guarantee the tanker owner's contractual liability to the IOPC Fund, subject only to the defences available to shipowners and insurers under 1992 Civil Liability under an amendment to the Memorandum of Understanding between the IOPC Fund and the International Group of P&I Clubs; and

(ix) shipowners and P & I Clubs would reserve the right to withdraw from the voluntary scheme if any essential element of the 1992 Conventions affecting tanker owner liabilities were to be amended[161].

However, it has yet not become clear whether the potential financial imbalance created by the Supplementary Fund should be resolved by voluntary industry solutions or whether this topic should be addressed by legislation. Also other elements in the international regime for oil pollution damage are still being considered. These considerations might ultimately result in explicit amendments to the 1992 Civil Liability and Fund Conventions.

4.4. Reaction of the EC

The 2003 Protocol could have entered into force on the sole condition that the EC Member States[162] ratified the Protocol, as the total amount of contributing oil received in the calendar year 2002 in the territories of the EC Member States, party to the

---

[161] 92FUND/WGR.3/14/7, Annex II; LEG/CONF.14/12.

[162] As of 7 September 2004 19 Member States of the (enlarged) European Union are State Parties to the 1992 Protocol to the Fund Convention (Belgium, Cyprus, Denmark, Germany, Greece, Finland, France, Ireland, Italy, Latvia, Lithuania, Malta, the Netherlands, Poland, Portugal, Slovenia, Spain, Sweden and the United Kingdom). Estonia will join by 6 August 2005. Luxembourg is only a State Party to the 1969 Civil Liability Convention. Austria, the Czech Republic, Hungary and Slovakia are neither a State Party to the 1969 Civil Liability Convention, the 1971 Fund Convention or the 1992 Protocols to the Conventions.

1992 Fund Convention, as reported on 31 December 2003, equals 582,290, 863 tons[163].

In light of the exclusive Community competence in matters regulated by the provisions on jurisdiction, recognition and enforcement contained in Articles 7 and 8 of the 2003 Protocol (see above), the European Commission asked the Council for a mandate to negotiate certain parts of the Supplementary Protocol. The proposed mandate included the negotiation of an article providing for the possibility for the Community to become a party to the Protocol. In addition, there was, in the view of the European Commission, *a need for a mechanism ensuring a continued possibility to apply the Community rules on the recognition and enforcement of judgments. By contrast, the provisions on jurisdictions were not considered necessary to revise in the Supplementary Fund Protocol.* However, the European Commission added, *should the 1992 Civil Liability and Fund Conventions be re-opened for revision, which is not an unlikely scenario for the coming years, there is a need to analyse in great detail the compatibility between Regulation 44/2001 and the international oil pollution compensation regime and the reasons therefore.* A negotiation mandate relating to the Supplementary Fund Protocol was adopted on 19 December 2002 by the Justice and Home Affairs Council. This mandate did not, as the Commission had proposed, include the negotiation of an accession clause for the Community, but it did hold that: "*the Commission should ensure that the regime for recognition and enforcement of judgments, as laid down by Regulation 44/2001, shall continue to apply when judgements given by a court of a Member State are recognized and enforced in another Member State. In this respect the necessary contacts with third States should be taken by the Commission well in advance of the Diplomatic Conference 12-16 May 2003.*"[164].

The European Community has no competence to oblige its member states to ratify international conventions, so the decision to adhere to the voluntary third-tier regime depends solely on national souvereignty. Moreover, the text of the Supplementary Protocol does not allow for the European Community as such to conclude the Protocol, to be ratified, accepted, approved or acceded to only by States that are a party to the 1992 Fund Convention. However, "*in order to safeguard the Community interests in view of its external competence, while at the same time enabling the Member States to ratify the Protocol*", a Council Decision

---

[163] Amount based on contributing oil from Italy, the Netherlands, France, United Kingdom, Spain, Germany, Greece, Sweden, Portugal, Finland, Belgium, Denmark, Ireland, Cyprus and Malta. Poland received 779,271 tons of contributing oil to the 1992 Fund Convention, thus deemed to be a minimum receipt of 1 million tons of contributing oil in that Contracting State. Both Slovenia and Latvia reported a "nil return" to the 1992 IOPC Fund because the amount of oil received by these States was less than 150,000 tonnes, *i.e.* the Member State does not qualify to pay contributions to the 1992 IOPC Fund. (INTERNATIONAL OIL POLLUTION COMPENSATION FUND (2004), *o.c.*, 155).

[164] Proposal for a Council Decision authorizing the Member States to sign, ratify or accede to, in the interest of the European Community, the Protocol of 2003 to the International Convention on the Establishment of an International Fund for Compensation for Oil Pollution Damage, 1992, and authorizing Austria and Luxemburg, in the interest of the European Community, to accede to the underlying instruments, COM (2003) 534 final, 8 September 2003, 6-7.

has been taken to exceptionally authorize the Member States, with the exception of Denmark, to sign and ratify the Supplementary Fund Protocol in the interest of the Community[165].

In taking the decision it has also been highlighted that two Member States, Luxembourg and Austria, are not yet Contracting Parties to the underlying 1992 Civil Liability and Fund Conventions, which is a precondition for concluding the Supplementary Fund Protocol. The underlying Civil Liability and Fund conventions raise similar issues with respect to the relationship to Regulation 44/2001. Therefore, the decision also includes an "authorization" for these two Member States to accede to the 1992 Civil Liability and Fund Conventions[166].

It thus becomes clear how the exclusive competence of the European Community for matters relating to the jurisdiction, recognition and enforcement of judgments in civil and commercial matters, as regulated by Regulation 44/2001, influences the conclusion of international liability and compensation agreements in the sphere of maritime transport by the Member States of the European Union[167].

Following the adoption of the 2003 Supplementary Fund Protocol and its recent entry into face, no further action has been undertaken at Community level for the adoption of the COPE Fund Proposal.

5. CONCLUSIONS

We can conclude that oil pollution incidents, national regulations and the interpretation of the conventions by domestic judges and the pressure of a

---

[165] Council Decision 2004/246/EC of 2 March 2004 authorizing the Member States to sign, ratify or accede to, in the interest of the European Community, the Protocol of 2003 to the International Convention on the Establishment of an International Fund for Compensation for Oil Pollution Damage, 1992, and authorizing Austria and Luxembourg, in the interest of the European Community, to accede to the underlying instruments, *OJ* L 78/22, 16 March 2004.

[166] The text of the decision has been taken before to adhesion of ten new Member States to the European Union as at 1 May 2004 (Cyprus, Estonia, Hungary, Latvia, Lithuania, Malta, Poland, Slovakia, Slovenia and the Czech Republic) so the situation after the enlargement of the European Union has yet to be considered.

[167] It is noteworthy to recall the proposal of the European Commission to set up an obligatory consultation procedure on relations between Member States and third countries in shipping matters and on action relating to such matters in international organizations and an authorization procedure for agreements concerning maritime transport. According to this proposal the Member States and the Commission shall consult each other, in accordance with the procedures laid down in this Decision: (i) on questions concerning shipping matters and dealt with in international organizations; and (ii) on relations with third countries in shipping matters, and on the negotiation, conclusion and functioning of bilateral or multilateral agreements in this sphere (Article 1 Proposal for a Council Decision setting up a consultation procedure on ralations between Member States and third countries in shipping matters and on action relating to such matters in international organizations and an authorization procedure for agreements concerning maritime transport, COM (1996) 707 final, *OJ* C 113, 11 April 1997).

transnational organization, such as the European Community, can exert a strong influence on the development of the international oil pollution damage regime.

By introducing the COPE Fund Proposal, the European Commission has urged the international community to adopt the 2003 Supplementary Fund Protocol, thus providing for a (voluntary) third tier of compensation for oil pollution damage at global level. Through the voices of the EC Member States – parties to the 1992 Civil Liability and Fund Conventions – the main elements of the (initial) COPE Fund Proposal have been set at the international legal scenery. Furthermore, the European Community is exercising its exclusive competence in the jurisdiction, recognition and enforcement of judgments in civil and commercial matters to stimulate the EC Member States to ratify the 2003 Protocol.

For the time being the European Community has succeeded in its attempts to change the international regime by exerting high pressure on the international level. Nevertheless, it is questionable if the European Commission – having full membership of the IMO in view - will limit its efforts to this effect, so the question mark remains as to the reaction of the international level on this continuing pressure.

REFERENCES

- BELLAYER-ROILLE, A. (2000), *Le transport maritime et les politiques de sécurité de l'Union européenne*, Editions Apogée, Rennes, 474 p.
- DAMIEN, M.-M. (1999), *La politique européenne des transports*, Presses Universitaires de France, Paris, 127 p.
- FARANTOURIS, N.E. (2003), *European integration & Maritime transport*, Ant. N. Sakkoulas / Bruylant, Athens / Brussels, 465 p.
- GONSAELES, G. & MAES, F. (2002), Taak V. Het ontwikkelen van technische en juridische procedures met het oog op de evaluatie van degradatie aan het mariene milieu en het financieel verhaal op de vervuiler: Het internationaal aansprakelijkheidsregime voor schade bij olieverontreiniging op zee: onderzoek naar de vergoedbaarheid van milieuschade, in: MAES, F. (Ed.), *MARE-DASM "Duurzaam Beheer van de Noordzee" – Plan voor wetenschappelijk ondersteuning van een beleid gericht op duurzame ontwikkeling, studie in opdracht van de Diensten van de Eerste Minister, Federale Diensten voor Wetenschappelijke, Technische en Culturele Aangelegenheden*, Ghent, 143 p.
- INTERNATIONAL OIL POLLUTION COMPENSATION FUND (2002), *Report on the Activities of the International Oil Pollution Compensation Fund in 2001*, London, 192 p.
- INTERNATIONAL OIL POLLUTION COMPENSATION FUND (2003), *Report on the Activities of the International Oil Pollution Compensation Fund in 2002*, London, 180 p.
- INTERNATIONAL OIL POLLUTION COMPENSATION FUND (2004), *Report on the Activities of the International Oil Pollution Compensation Fund in 2003*, London, 184 p.
- LEIGHTON, B. (1996), *EC Transport Law – A Practitioner's Guide*, FT Financial Publishing, London, 220 p.

- LENAERTS, K. & VAN NUFFEL, P. (2003), *Europees recht in hoofdlijnen*, Maklu, Antwerp, 900 p.

- PALLIS, A.A. (2002), *The Common EU Maritime Transport Policy – Policy Europearisation in the 1990s*, Ashgate Publishing Ltd., Hampshire, 283 p.

- POWER, V. (1992), *EC Shipping Law*, Lloyd's of London Press Ltd., London, 771 p.

- STEVENS, H. (2004), *Transport Policy in the European Union*, The European Union Series (Palgrave MacMillan), Hampshire/New York, 276 p.

- UNION DES AVOCATS EUROPEENS (1999), *L'Europe des Transports – Régulation, Dérégulation, Impact du passage à l'Euro – XIIe Congrès de l'Union des Avocats Européens (Marseille, 8-9 octobre 1998)*, Bruylant, Brussels, 335 p.

PART III.

ECONOMIC VALUATION OF DAMAGE TO THE MARINE ENVIRONMENT

CHAPTER 7

ECONOMIC ASSESSMENT OF MARINE QUALITY BENEFITS: APPLYING THE USE OF NON-MARKET VALUATION METHODS

PAULO A.L.D. NUNES

Department of Economics
University of Venice
Fondazione Eni Enrico Mattei (FEEM)

Pallazo Querini Stampalia
30122 Venice, Italy
E-mail: pnunes@unive.it

and

ARIANNE T. DE BLAEIJ

Faculty of Geosciences
Urban and Regional Research Centre Utrecht (URU)
Utrecht University

PO Box 80115, 3508 TC Utrecht, The Netherlands
E-mail: A.deBlaeij@geog.uu.nl

1. INTRODUCTION

In a democratic system, policy makers should take account of the preferences of the taxpayers belonging to that system. Because we live in a world with scarce resources, one is asked to make the choice regarding the use and management of

F. Maes (ed.), Marine Resource Damage Assessment, Liability and Compensation for Environmental Damage, 135–163.

these resources. In this context, if policy makers decide to invest in the protection of marine ecosystems, fewer financial resources would be available for other policy areas, such as national defence. In addition, the investment in the protection of marine ecosystems brings along with it the provision of public values, which are fully priced on current markets. In other words, marine ecosystems provide a wide range of benefits to humans and most are not valued on market prices. For example, marine ecosystems provide an important role in balancing the local chemical composition of the water and we do not observe a market price that reflects such benefit. Given that most human activities are priced in one way or other, in some decision contexts, the temptation exists to downplay or ignore marine ecosystem quality benefits on the basis of non-existence of prices for marine benefits such as local marine biodiversity. The simple and simplistic idea here is that a lack of prices is identical to a lack of values. Clearly, this is a slightly biased perspective.
The micro-economic theory of externalities teaches us that many values cannot be incorporated in conventional market transactions. The question then is how to translate such values into monetary dimensions. This is a challenging question to be addressed by economists. The underlying idea is that economists need to rely on particular economic valuation methods in order to retrieve the monetary value of these marine benefits. Since these are not directly observed in the market, the valuation methods are called non-market valuation methods and constitute the core of the present chapter. We will articulate the discussion as follows. Section 2 presents and discusses the motivations to perform economic valuation, in general, and non-market valuation of marine ecosystem quality, in particular. Section 3 provides a discussion regarding the link of non-market valuation and microeconomic theory and consumer behaviour. Section 4 illustrates the wide range of different valuation methodologies and its link to the different economic value components of marine environmental quality. Section 5 examines in detail two economic valuation applications, illustrating their crucial role for policy guidance. Finally, we will provide some conclusions.

2. MOTIVATION FOR ECONOMIC VALUATION

2.1. Introduction

The economic valuation of environmental assets in general and marine resources in particular is among the most pressing and challenging issues confronting environmental economics. Major organizations across the world such as the World Bank[1], Resources for the Future[2] and OECD[3] and US governmental agencies like the

---

[1] WANG, H, & WHITTINGTON, D. (2000), *Willingness to Pay for Air Quality in Sofia, Bulgaria*, World Bank Policy Research Working Paper, n. 2280, Washington D.C., US; WANG, H., LAPLANTE, B., WU, X., & MEISNER, C. (2004), *Estimating Willingness-to-Pay with Random Valuation Models: An Application to Lake Sevan, Armenia*, World Bank Policy Research Working Paper, n. 3367, Washington D.C., US.

[2] MITCHELL, R.C., & CARSON, R.T. (1989), *Using Surveys to Value Public Goods: The Contingent Valuation Method*, Resources for the Future, Washington DC, US.; KRUPNICK, A., ALBERINI, CROPPER, M., SIMON, N., O'BRIEN, B., GOEREE, R., & HEINTZELMAN, M. (2000), *Age, Health,*

Environmental Protection Agency[4] carry out economic value assessment applications. One may wonder about the reason why such monetary assessments of environmental resources are undertaken. Four main reasons can be identified. These are performing cost-benefit analysis, environmental accounting, assessing natural resource damage, and carrying out proper pricing. These will subsequently be considered in more detail.

2.2. Cost-benefit analysis

Cost-benefit analysis (CBA) is a welfare-theoretic method to trade-off the advantageous and disadvantageous effects of a proposed project by measuring them in monetary terms. CBA emerged as an attempt to systematically incorporate economic information that can be applied to project and policy evaluations. Since CBA has traditionally been defined in terms of gains and losses to society, project-oriented CBA has tended to be confined to public sector investment projects. The first evaluation studies were carried out in the US in the 1950s to deal with 'intangibles' in a consistent way, *e.g.*, for river basin projects and infrastructure projects. These methods found much application, inter alia in World Bank practices. They were also heavily criticized for many inherent shortcomings, which has led to many new or adjusted methods, such as cost-effectiveness analysis, goals-achievement methods and multicriteria analysis[5].
The use of CBA to evaluate policy is more recent[6]. Like an investment project, policies have costs and benefits. For example, standards for marine pollutants concentrations and taxation of marine pollutants are two different policies, which, in turn, are associated with different gains and losses to society. The basic rule of CBA in decision-making is to approve any potentially worthwhile policy if the benefits of the policy exceed the costs. Moreover, to make the best choice, a decision-maker should opt for the policy option with the greatest positive net present value. Other criteria exist, such as ranking and evaluating projects according to their 'internal rate of value' or according to the 'benefit cost ratio'[7].
From the policy agenda point of view, CBA has been used in the US for evaluating policies since the late 1970s. However, only after REAGAN's Executive Order

---

and the Willingness to Pay for Mortality Risk Reductions: A Contingent Valuation Survey of Ontario Residents, Resource for the Future Discussion Paper, n. 00–37, Washington D.C., US.

[3] BILLER, D., & BARK, B. (2001), *Valuation of Biodiversity: Selected Studies*, OECD, Environment Directorate, Paris, France.

[4] MANSFIELD, C., VAN HOUTVEN, G., JOHNSON, F.R., CRAWFORD-BROWN, D., & ZACHARY, P. (2002), *Behavioral Reactions to Ozone Alerts: What Do They Tell Us About Willingness-to-Pay for Children's Health?*, Valuation of Environmental Impacts on Children's Health, EPA Research Project, Washington D.C., US.

[5] See NIJKAMP, P., RIETVELD, P., & VOOGD, H. (1991), *Multicriteria Evaluation in Physical Planning*, North-Holland, Amsterdam, The Netherlands.

[6] See for an overview BOARDMAN, A., GREENBERG, D., VINING, A. , & WEIMER, D. (2000), *Cost Benefit Analyis: Concepts and Practice*, 2nd Edition, Prentice Hall.

[7] See for a literature review on CBA and its application to environmental issues, HANLEY, N., & SPASH, C.L. (1993), *Cost-Benefit Analysis and the Environment*, Edward Elgar, Aldershot, UK and Brookfield, US.

12291, in 1981, has CBA been extensively used for evaluating new regulations. In contrast, in Europe there are no legal requirements for CBA for new regulations. An exception is the UK, where the 1995 Environment Act envisions the use of CBA in policy making. Clearly, the use of and the critical judgments of CBA in public policy is still a matter of ongoing scientific debate among economists.

2.3. Environmental accounting

Various efforts have been made to adjust national accounting systems and associated gross national product (GNP) statistics for the depreciation of environmental assets and for negative externalities such as pollution and the loss of biodiversity. The theoretical literature explores alternative ways of adjusting conventional estimates of national income to reflect environmental deterioration[8]. Green accounting is one possible strategy.

The underlying idea is to add to the traditional national accounting system information on physical flows and stocks of environmental goods and services – the so-called 'physical satellite accounts'. In the Dutch context, for example, the Netherlands Central Bureau for Statistics developed the NAMEA, a National Accounting Matrix that includes both Economic and Environmental Accounts[9]. An important aim of green accounting is to obtain an adjusted 'green' GNP. This can play a potentially crucial role in policy making since the GNP has a powerful influence on macro-economic policy, financial markets and international institutions (OECD, IMF, and World Bank). If national income is wrongly estimated, then economic analysis and policy formulation are based on the wrong premises, thus 'steering' the society by the wrong compass[10]. Adjustment of the national accounts to reflect marine ecosystem quality loss will lower the GNP[11].

Nevertheless, practice shows that the adjustment of national accounting systems is not an easy task. It is therefore necessary to achieve international agreement about harmonizing GNP adjustments, allowing for the comparison of GNP and national accounts between countries. Independent of which valuation methods are used for this purpose, it is clear that monetary valuation of the depreciation of environmental assets and negative externalities, such as pollution and the loss of biodiversity, is a key element in green environmental accounting.

---

[8] ARONSSON, T., JOHANSSON, P.-O. & LOFGREN, K.-G. (1997), *Welfare Measurement, Sustainability and Green' Accounting - a Growth Theoretical Approach*, Edward Elgar, Cheltenham, UK and Lyme, US.

[9] KEUNING, S.T., & DE HAAN, M. (1996), *What is a NAMEA?*, Statistics Netherlands National Accounts Paper, Voorburg, The Netherlands.

[10] HUETING, R. (1980), *New Scarcity and Economic Growth*, North-Holland, Amsterdam, The Netherlands; EL SERAFY, S. (1999), *Steering the Right Compass: Proper Measurements for Sound Macro-economic Management*, paper presented at International Symposium Valuation of Nature and Environment, Royal Netherlands Academy of Arts and Sciences, Amsterdam, The Netherlands.

[11] GERLAGH, R., DELLINK, R., HOFKES, M., & VERBRUGGEN, H. (2002), A Mesure of Sustainable Development for the Netherlands, *Ecological Economics*, 41, 157–74.

2.4. Natural resource damage assessment and legal claims

Natural resource damage assessments (NRDAs) appraise how much society values the destruction of natural resources. An important benchmark in the history of NRDA is the massive oil spill due to the grounding of the oil tanker *Exxon Valdez* in Prince William Sound in the northern part of the Gulf of Alaska on 24 March, 1989. This was the largest oil spill from a tanker in US history. More than 1,300 km of coastline were affected and almost 23,000 birds were killed[12]. After the oil spill, the State of Alaska commissioned a legal action in order to assess EXXON's financial liability in the damage to the natural resources. A national contingent valuation study estimated the loss to US citizens as a result of the oil spill. The natural resource damage resulting from the *Exxon Valdez* oil spill was estimated at $ 2.8 billion. For the first time, a governmental decision expressed the legitimacy of nonuse values as a component of the total damage value. To date, NRDAs are only undertaken in the US and have not yet become an issue in the European policy agenda because of different legal arrangements.

3. THE ECONOMIC VALUATION PERSPECTIVE

3.1. Introduction

Economic analysis and the valuation of marine ecosystem quality is based on an instrumental perspective on the value of biodiversity. This means that the value of marine ecosystem quality is regarded as the result of an interaction between humans and the object of valuation, which is 'changes in marine ecosystem quality'. 'Economic value' does not denote an absolute value of levels, but of system changes, preferably marginal or small ones. The reason for this is that the theoretical basis of economic valuation is monetary (income) variation as the response to a certain policy or environmental change. Therefore, the terms 'economic value' and 'welfare change' can, in principle, be used interchangeably. Therefore, economic valuation provides a monetary indicator of marine ecosystem quality values. The reason for this is that the theoretical basis of economic valuation is monetary (income) variation as a compensation or equivalent for direct and indirect impact(s) of a certain marine ecosystem quality change on the welfare of humans. Explicit marine ecosystem quality changes, preferably in terms of accurate indicators, should be related to these. The economic valuation of marine ecosystem quality changes is based on a reductionist approach value. This means that the total economic value is regarded as the result of aggregating various use and nonuse values, reflecting a variety of human motivations, as well as aggregating local values to attain a global value, *i.e.* a bottom-up approach[13]. Moreover, the economic valuation of marine

[12] CARSON, R.T., MITCHELL, R.C. , HANEMANN, W. M., KOPP, R.J., PRESSER, S., & RUUD, P. A. (1992), *A Contingent Valuation Study of Lost Passive Use Values Resulting from the Exxon Valdez Oil Spill*, Report prepared for the Attorney General of the State of Alaska, Washington, US.

[13] See NUNES, P.A.L.D., & VAN DEN BERGH, J.C.J.M. (2001), Economic Valuation of Biodiversity: Sense or Nonsense, *Ecological Economics*, 39, 203–22; NUNES, P.A.L.D. (2002), *The Contingent Valuation of Natural Parks: Assessing the Warm Glow Propensity Factor*, New Horizons in

ecosystem quality starts from the premise that social values should be based on individual values, independently of whether the individuals are experts in biodiversity-related issues or not. This can be considered consistent with the democratic support of policies.

3.2. The concept of economic value

A basic micro-economic premise is that individuals make welfare-optimizing consumption decisions. These decisions are captured in the consumer demand functions with respect to available goods and services. Marine ecosystem quality considerations enter into these demands. To illustrate this setting, we consider an individual whose utility function has the following form:

$$V = V(x, q, z)$$

Here x is the consumption of the private good, q the quantity of the environmental resource, and z a marine ecosystem quality indicator. For example, q could represent the number of recreational sites and z the level of marine species richness. We assume that x is a composite private good whose price is normalized to one. p is the price associated with q. This framework allows the study of a welfare change in the marine ecosystem quality indicator, z. This change may be interpreted as the introduction of a set of new regulations designed to allow commercial development in protected areas.

In the original situation, *i.e.* before the implementation of the new regulation, the individual faces a particular marine ecosystem quality level. Let us denote such a level by z^0. For a marine ecosystem quality level z^0, and given the consumer monetary income M, the consumer maximizes $V(x,q,z)$. This yields an optimal consumption bundle (x^0, q^0), with $q^0 (p, M, z^0)$ and $x^0 (p, M, z^0)$ and a utility level $V^0(x^0, q^0, z^0)$. Inserting the demand functions into the utility function gives the indirect utility function $V(x^0 (p, M, z^0), q^0 (p, M, z^0, z^0) = v(p, M, z^0)$.

Table 1 summarizes the notation.

Table 1. *Summary of the notation*

Variables and function of interest	*Original situation*	*New situation*
Marine ecosystem quality level	z^0	z^1
Utility level	V^0	V^1 with $V^0 > V^1$
Indirect utility function	$v(p, M, z^0)$	$v(p, M, z^1)$

Environmental Economics Series, Edward Elgar, Cheltenham, UK and Northampton, MA, US; NUNES, P.A.L.D., & SCHOKKAERT, E. (2003), Identifying The Warm Glow Effect In Contingent Valuation, *Journal of Environmental Economics and Management*, 45, 231-45.

The literature suggests two alternative measures that can be used to assess the magnitude of the welfare change as described by the introduction of the new regulation. These are the Hicksian compensating measure and the Hicksian equivalent measure, which are theoretical refinements of the ordinary consumer surplus[14]. The Hicksian compensating welfare measure equals the compensating payment, *i.e.* an offsetting change in income to make the individual indifferent to distinctions between the original situation (status quo) and the new situation. The Hicksian compensating variation *(HC)* is the solution to

$$v\left(p, M, z^{0}\right) = v\left(p, M + HC, z^{1}\right) = V^{0}$$

i.e. the *HC* measures what must be paid to the individual to make that person indifferent to the new environmental quality level. In other words, under the new situation, the individual's income would be increased by the amount of *HC*, but the person would still be as well off as in the original situation.
The Hicksian equivalent welfare measure corresponds to change in income that would lead to the same utility change as the new situation. The Hicksian equivalent *(HE)* is the solution to

$$v\left(p, M, z^{1}\right) = v\left(p, M - HE, z^{0}\right) = V^{1}$$

i.e. the *HE* measures the income change that is equivalent to the welfare lost due to the new situation. In other words, that income change is the 'price' that reflects the consumer's maximum willingness to pay (WTP) to avoid an undesirable change in *z*. This interpretation assumes that the benchmark is the level of welfare after the change. If, however, the changes are being compared with the initial situation, then we measure the willingness to accept (WAC). In other words, the two alternative Hicksian welfare measures can be interpreted in terms of the implicit rights and obligations associated with alternative environmental quality levels. In this context, the *HC* carries with it implicitly the assumption that the individual has the use, property and freedom related to the original environmental quality. In contrast, the *HE* is consistent with the idea that the individual has an obligation to accept a reduction in environmental quality. The choice between them is, therefore, ultimately an ethical one. In short, it reflects a value judgment about which underlying distribution of property rights is more equitable[15]. Table 2 summarizes the results and the preferred welfare measure, according to the suggestions of the NOAA panel[16].

---

[14] HICKS, 1943.
[15] KRUTILLA, 1967.
[16] NOAA, 1993.

Table 2. *Hicksian welfare measures and the property rights distribution*

Attribute quality	*Hicksian equivalent measure (Implied property rights in the change)*	*Hicksian compensating measure (Implied property rights in the original situation)*
Increase	WAC to forgo	WTP to obtain
Decrease	WTP to avoid	WAC to accept

Different valuation methods are available to compute a monetary estimate with respect to the WTP and WTA welfare measures. These will be presented and discussed in detail in the next session.

4. MONETARY VALUATION METHODS

4.1. Introduction

In a conceptual framework, one can define the total value (TV) of the marine quality benefits in terms of the use value (UV) and nonuse value (NU) related to the protection of marine resources. The former can be further divided into direct and indirect use values (DUV and IUV). Direct use values include: (a) marine tourism and coastal recreation benefits; (b) natural and cultured marine species with commercial value; and (c) insurance with respect to potential risks to human health – see Table 3.

Indirect use values refer to benefits that relate to the well functioning of the marine ecosystem and the survival of marine living resources, even if these have no direct commercial value. Finally, nonuse values of marine quality can be divided into a bequest value (BV) and an existence value (EV). Bequest value refers to the benefit accruing to any individual from the knowledge that future generations might benefit from a sustainable marine ecosystem. Existence value refers to the benefit derived simply from the knowledge that the marine ecosystem is protected without even being used. Both direct and indirect values, relating to production, consumption and nonuse values of marine ecosystem quality are considered when assessing the economic value of marine ecosystem quality. Bearing in mind the particular nature of each benefit under consideration, the economist can use a wide range of valuation methods including travel cost and stated preference methodologies. These will be presented and discussed in the following subsection.

Table 3. *Classification of economic values provided by the protection of marine quality*

Value components		*Examples of benefits*	*Most suitable valuation technique*
Use value (UV)	Direct use value (DUV)	Tourism and recreational benefits, *e.g.* visits to the beach, swimming and sailing	Travel cost method
		Marine resources with commercial value, *e.g.* fish, shellfish and mollusk	Aggregate price analysis*
		Human health, *e.g.* prevention of skin allergies and gastrointestinal disorders	Stated preference methods
	Indirect use value (DUV)	Marine ecosystem and ecological functioning, *e.g.* balancing local chemical composition of the water, protection of local marine living resources diversity	Stated preference methods
Nonuse value (NUV)	Bequest Value (BV)	Legacy benefits, *e.g.* heritage of marine living resources for future generations	Stated preference methods
	Existence value (EV)	Existence benefits, *e.g.* knowledge guarantee that some marine living resources are not extinct	Stated preference methods

Note: * Market price valuation technique

4.2. Economic valuation methods

An accurate, complete and reliable monetary assessment of marine ecosystem benefits requires the application of specific monetary valuation tools. Bearing in mind the classification of the economic value component and respective marine benefit, Table 3 shows the most suitable valuation methodology to be used. As one can see, the travel cost method is the most suitable valuation method for monetary value assessment of marine quality benefits that relate to the provision of tourism and recreational opportunities. In addition, stated preference methods can fulfill an important role in the overall assessment of marine quality benefits. Indeed, stated preference methods can be applied to assess the monetary value of most of the types of benefits provided by the protection of marine quality, including both the bequest and existence values. Furthermore, these methods have the advantage that marine policies may be valued even if they have not yet been adopted (*ex ante* valuation) or lie outside the current institutional arrangements. Thus, they offer much scope and flexibility for specifying different marine protection, restoration and amelioration programmes. For these reasons we will pay particular attention to the analysis and discussion of stated preference methods, including the stated choice method (SC) and contingent valuation method (CV). In addition, we will also focus on the use of the travel cost (TC) method to assess marine recreational values[17].

SC and CV are survey based valuation techniques that are widely used in the context of environmental valuation[18]. CV is a survey-based approach that directly estimates the preferences for risk reductions in the overall marine ecosystem quality. Therefore, CV gives an immediate monetary estimate of the WTP welfare measure associated to an increase in the marine quality. In short, CV makes use of a questionnaire that describes a survey market in which non-market goods can be traded. It is assumed that the values elicited with CV will correspond with those that would emerge on real markets. The contingent market defines the good itself, the institutional context in which it would be provided, and the way it would be financed. Respondents are then asked to express their maximum WTP for a survey described change in the level of the environmental good. Alternatively, the SC approach confronts respondents with a set of two or more survey described alternatives that differ in terms of the respective attributes and attribute levels. The respondents are asked to choose their preferred alternative. The non-market valuation method is also known in the literature as 'choice modeling' or 'conjoint choice'. When using SC the economist often uses the status quo among the two alternatives that the respondent is asked to choose. Such an alternative is necessary in order to be able to interpret the results, namely to study the individual WTP for the proposed marine quality improvement and its survey described attributes[19]. Both a CV and a SC questionnaire contain two important elements. The first is a clear description of the environmental good to be valued. In case of CV this is the status quo, and the proposed changes. For SC these are the attributes of the alternatives and the related attribute levels. The second element is a mechanism for eliciting the WTP of the respondents. CV describes the elicitation process in the question, while in SC, the elicitation process is one of the attributes of the alternatives. Finally, one can make use of the travel cost method (TC) when the valuation exercise, and underlying policy proposals, refer exclusively to the recreational use values provided by

---

[17] A third method mention in Table 3, aggregate price analysis, will not be discussed since this valuation approach is based on market prices falls and not on the use of non-market valuation methods.

[18] NOAA 1993 ; CARSON, R.T., MITCHELL, R.C. , HANEMANN, W.M., KOPP, R.J., PRESSER, S., & RUUD, P.A. (1992), *o.c.*; and HANLEY, N., MOURATO, S. *et al.* (2001), Choice modelling approaches: A superior alternative for environmental valuation?, *Journal of Economic Surveys* 15(3); 435-462.

[19] Quoting BOYLE *et al.* (2001): "*If the choices do not include a status quo alternative, a nonzero value is implied in the estimated likelihood function for people who would not choose any of the alternatives. In general, this serves to bias estimates upwards.*"

protection of marine quality. In short, the travel-cost method is a demand-based model for use of a recreation site of sites. The model is based on data about trips made by the sample population (see LOOMIS book). Two case studies regarding the application of the CV, SC and TC models in the domain of marine environmental quality will be presented in the following section.

5. CASE STUDIES

5.1. Introduction

In this section we review two marine ecosystem valuation studies carried out in the Netherlands. The aim is to show the relevance of this type of information for cost-benefit analysis and respective policy guidance implications. The first study presented here is a joint travel cost-contingent valuation study to value the protection of the Dutch coastline from biological pollution events. The second study examines the use of the stated choice valuation method to rank alternative cockle fishery management practices in the Dutch Wadden Sea bearing in mind their respective marine environmental effects[20].

5.2. Study 1: ballast reception facility in the harbor of Rotterdam[21]

This study performs a joint travel cost – contingent valuation exercise to assess the economic value of a marine policy programme targeted at the protection of the coastline of the Netherlands from biological pollution events, including harmful algal blooms (HABs). The term ‘harmful’ covers a set of algae species that share one characteristic: they have a negative impact (causing damage) on marine living resources and ecosystems as well as on humans. In general terms, three categories of negative effects can be associated with HABs: risks to human health, constraints on tourism and recreation, and bio-ecological effects. The latter includes the so-called red tides, which are responsible for the destruction of marine living resources[22]. The most common pathway through which (harmful) micro-algae are introduced is through the uptake, transport and subsequent discharge of ballast water and associated sediments by ships. Introductions of harmful algae by ballast water may, for instance, be prevented through: (i) ballast water exchange at sea; (ii) non-or minimal release of ballast water in both coastal waters and ports; (iii) discharge of ballast water to appropriate reception facilities. The latter involves the use of alternative, and often expensive, technologies for the treatment of ballast water. By contrast, the welfare gains associated with the introduction of such an

---

[20] The reader can find an additional contingent valuation study that focus on the monetary value assessment regarding a national protection plan against oil spils in the Belgian coast line (VAN BIERVLIET, K., LE ROY D. & NUNES, P.A.L.D., *A contingent valuation study of an accidental oil spill along the Belgian coast* (Part III – Chapter 8).

[21] For a detailed information see NUNES, P.A.L.D., & VAN DEN BERGH, J.C.J.M. (2004), Can People Value Protection against Invasive Marine Species? Evidence from a Joint TC–CV Survey in the Netherlands, *Environmental and Resource Economics*, 28, 517–32.

[22] See VAN DEN BERGH, J.C.J.M. *et al.* (2002).

infrastructure is often unknown since most of the benefits have non-market characteristics. For this same reason, the present study proposes to evaluate the benefits connected with the introduction of a ballast water treatment complex in the harbour of Rotterdam, the Netherlands.
The valuation exercise presents distinctive features that justify particular attention. We refer to: (i) the North Sea in general, and the North-Holland coastal area in particular, have seen relatively few applications of economic value assessment of marine-coastal management programmes; (ii) the use of alternative economic valuation methodologies in accordance to their degree of suitability with respect to the multifaceted nature of the non-market benefits generated by the introduction of ballast water complex; (iii) the management of different qualitative and quantitative methodological settings and analyse their impact on the economic value of the different non-market benefits under consideration; and (iv) the integration of the economic estimation results in cost-benefit analysis and thus create a platform for the design of an efficient ballast water policy.

5.2.1. Survey instrument and survey execution

The administration of the questionnaire took place during August 2001 in *Zandvoort*. This is the most popular beach resort in the Netherlands and it is located in the province of North-Holland. *Zandvoort* offers beach recreation activities, ranging from swimming, cycling, kiting, windsurfing to sailing. Using a simple random sample selection mechanism, the interviewer contacted 352 groups of visitors, 242 of whom completed the questionnaire. The participation rate is therefore 69 percent. Unlike many other survey applications the most often mentioned reason for refusal is "not speaking Dutch". Tourists from abroad accounted for one third of all refusals[23]. The questionnaire's demographics and socio-economic characteristics indicate that the median respondent is 41 years old and has a household monthly net income in the range of 1,800 € – 2,300 €. The majority of the respondents were women. When confronting the data of our survey with socio-demographic statistics for the Netherlands we were unable to find major differences. In other words, the sample is quite representative of the Netherlands.
Two major parts characterize the structure of the questionnaire. The first half of the questionnaire collects data on the respondents' travel. This information will be used to shed light on the recreation-related benefits of such marine protection programme. The second half of the questionnaire retrieves information for undertaking a contingent valuation exercise. The contingent valuation exercise was designed to assess the economic vale of the non-market benefits provided by the implementation of a ballast waters complex in the Netherlands. The benefits of such a protection programme are described in terms of its bio-ecological effects, including the protection from the so-called red tides, which are responsible for the destruction of marine living resources. Since respondents will be unfamiliar with such a market,

[23] The relatively high overall nonresponse rate needs to be examined carefully. In fact, from the policy perspective, the relevant population is the domestic one, *i.e.* those who are in a position to be taxed by the Dutch national government. Having said that, the response rises to 81 percent.

the ballast water programme needs to be carefully described. This occurs in terms of: (i) construction of a ballast water disposal treatment complex in the Rotterdam harbor, an internal circuit where ballast water will be transferred to and submitted to an appropriate physical-chemical treatment before being discharged into the sea; and (ii) implementation of an algae monitoring programme of the water quality in the open sea and along the North-Holland beaches. Finally, respondents were asked to state their opinion about alternative marine management policies. Coastal management activities related to 'protecting the marine ecosystem from water pollution' and 'ensuring seawater quality that does not provoke skin allergies' are associated with a high ranking index. Such a ranking pattern confirms the relevance of the present valuation study of damages caused by HABs.

5.2.2. Travel cost estimates

Survey responses on the travel data show that more than 50 percent of the respondents took a car to travel to *Zandvoort*, 20 percent arrived by public transport (*i.e.* train), 10.7 percent came by bicycle, and 15.6 percent of the respondent arrived on the beach on foot. We used public domain software of the Dutch Railways to computed travel time for public transport. The sample results show that beach visitors who traveled by train spent on average two hours and twenty-five minutes on the two-way journey[24]. Each respondent traveling by car was asked for the postal zip code of their address, the make and model of their car, the size of the engine, the type of fuel used, and the parking fees at the beach. These respondents traveled on average 104 kilometers and spent on average one hour and twenty-three minutes on the two-way journey. On the basis of this information we were able to compute individual car travel costs.

The generalized travel costs of a visit are defined as the sum of two components: transportation costs and travel time. The transportation cost is calculated as a function of the respondent's means of transport; in this case (i) car, (ii) train, (iii) bicycle and (iv) walking. For the latter two categories the transportation costs are assumed to be zero. The train costs are estimated for a two-way train ticket. For respondents traveling by car, the transportation cost is calculated according to the fuel used per kilometer. The time cost is estimated by multiplying the amount of time that a respondent spends on the two-way trip by the value of time. The value of time varies according to respondent's monthly income and selected means of transport. The different values reflect differences in consumers' preferences with respect to the choice of the means of travel, incorporating expected congestion time for car users.

Estimation results are presented in Table 4. This table shows the estimates of the partial derivatives regarding the vector of explanatory variables, *i.e.* the marginal effects. By adopting this specification we are able to provide key elasticities, which play a crucial role in terms of the respective interpretation in terms of policy

---

[24] The train station at *Zandvoort* is less than five minutes' walk from the beach. As the origin we used the closest train station to the zip code reported by the respondent. For respondents living far from a railway station we included an extra twenty minutes for transport from home to the railway station.

analysis. In this context, we can see that all the three variables that represent the generalized travel cost have a negative sign, reflecting that the higher the individual travel cost is, the lower is the number of annual visits to *Zandvoort*. Second, the transport expenditure parameter magnitude when compared to the remaining travel cost variables turns out to be a highest estimate and is the only magnitude statistically significant different from zero. This shows that the price elasticity of demand of trips, measured in terms of transport expenditures, is particularly relevant in explaining consumer recreational behavior, *i.e.* in determining the number of annual visits to the beach. Such information turns out to be of crucial information for policy guidance since any measure that directly targets the transport expenditures, including taxes on fuel, will influence the annual trips to *Zandvoort*. By contrast, according to Table 4, the prices of the parking space available at the beach does not change consumer recreational behavior since the respective parameter estimate is not statistically significant from zero. Therefore, any policy option that is characterized by changing the parking rates will not reduce he average number of annual visits to *Zandvoort* nor the total welfare of the visitors. In addition, Table 4 shows that an increase in the travel time would reduce the number of visitors to the beach of *Zandvoort*. However, the respective price elasticity is not statistically significant. Therefore, one can interpret this result as signal that the value of travel time savings is not particularly relevant to explain visitation behavior.

Furthermore, Table 4 shows that the demand of visits to the beach turns out not to be particularly sensitive to income, although the income elasticity is statistically different from zero. A one percent increase of the net household monthly income corresponds to an increase of 0.82 % in the total annual visits. With regard to personal characteristics, the estimation results show that the number of visits is expected to be lower for male, older respondents living with a partner than for other respondents. Furthermore, visitors who plan to stay all day long at the beach have a higher annual visit frequency than respondents who visit the beach for half-a-day or a couple of hours. Finally, estimation results indicate that respondents who visit the beach of *Bloemendaal*, which corresponds to the location of the *Kennemer* dunes natural park, have a higher annual visit frequency than respondents that visit the beaches of *Zandvoort*.

Table 4. *Travel cost estimation results* [(a)]

Variable	*Estimate*	*p-val.*
Intercept	4.8966*	0.00
Travel costs		
Transport	– 0.0139*	0.04
Travel time	– 0.0101	0.34
Parking	– 0.0015	0.91
Site characteristics		
Bloemendaal beach	0.8720**	0.11
Sunny weather	0.0724	0.85
Week-end	0.4847	0.25
Personal characteristics		
Male	– 0.3192	0.37
Age	– 0.0118	0.42
Net income (by the hundreds of €)	0.0082**	0.06
Field of studies: Economics	– 0.3865	0.28
Living with a partner	– 0.5590	0.20
Stay at the beach all day	0.7079**	0.09
Log-likelihood	– 280.65	
Annual gross recreation benefit per individual	55 €	

Notes: * (**) Statistically significant at 5% (15%)
(a) Marginal effects model. Full information maximum likelihood estimator with correction for sample selection is estimated with *LIMDEP*.

5.2.3. Contingent valuation estimates

To better mimic market behaviour, the present CV exercise uses a dichotomous-choice WTP question[25]. Following HANEMANN *et al.*[26], a follow-up valuation question was included so as to improve the statistical efficiency of the monetary estimates. All respondents that answered “no-no” faced a follow-up, open-ended WTP question. If the response to this was a zero willingness-to-pay, then the respondent is asked to indicate her major motivation for this choice. According to the double bounded response model, for each respondent *j* four possible response outcomes are possible: “no/no”, “no/yes”, “yes/no” and “yes/yes”, coded as r_{nn}^{j}, r_{ny}^{j}, r_{yn}^{j} and r_{yy}^{j} binary indicator variables, respectively. As one would expect from economic theory, the proportion of ‘yes-yes’ responses falls sharply with the amount the respondent is asked to pay. Only 3.8 percent of respondents state a willingness to

[25] CAMERON, T. A., & JAMES, M.D. (1987), Efficient estimation methods for closed-ended contingent valuation surveys, *American Journal of Agricultural Economics*, 269-276.

[26] HANEMANN, M.W., LOOMIS, J., & KANNINEN, B. (1991), Statistical Efficiency of Double-bounded Dichotomous Choice Contingent Valuation, *American Journal of Agricultural Economics*, 73(4), 1.255-1.263.

pay above 123 €. Such a low sample proportion at the highest bid indicates that the bid card has was well designed – see Table 5. In addition, the proportion of 'no-no' responses increases the bid amount. The remaining answering patterns, "yes-no" ("no-yes") responses, indicate that the respondent's maximum WTP lies between the initial bid amount and the increased (decreased) bid amounts.

Table 5. *Bid cards*

	Monetary amounts[*]			*Distribution of the WTP responses (in %)*			
Bid card	Initial	High	Low	Yes-yes	Yes-no	No-yes	No-no
n. 1	6.5 €	20 €	2.5 €	15.5	4.6	0.4	4.2
n. 2	14 €	34 €	7 €	11.7	6.7	0.8	5.9
n. 3	20 €	52 €	11 €	10.5	10.9	0.8	2.5
n. 4	40 €	123 €	16 €	3.8	9.2	5.9	6.7

Note: [*] Originally formulated in Dutch Guilders.

Bearing in mind the four possible response outcomes, the sum of contributions to the likelihood function $L(\theta)$ over the sample is maximized[27]:

$$L(\theta)=\sum_{j=1}^{N}\left\{\begin{array}{l} r_{nn}{}^{j}\ \ln F\left(b_{l}^{j};\theta\right)+r_{ny}{}^{j}\ \ln\left[F\left(b_{i}^{j};\theta\right)-F\left(b_{l}^{j};\theta\right)\right]+ \\ +r_{yn}{}^{j}\ \ln\left[F\left(b_{h}^{j};\theta\right)-F\left(b_{i}^{j};\theta\right)\right]+r_{yy}{}^{j}\ \ln\left[1-F\left(b_{h}^{j};\theta\right)\right]\end{array}\right\}$$

where $F(.)$ is a statistical distribution function with parameter vector $\theta \equiv (\beta, \sigma)$, where β and σ denote the location and scale parameters of the distribution. For a univariate model with a lognormal distribution, the mean WTP is given by $WTP = e^{\beta + 0.5 \times \sigma^2}$, where β and σ represent the location and scale parameters. As we can see in Table 6, the β and σ standard errors indicate that the parameters are estimated precisely. The annual mean WTP estimate is 76.2 €.

Table 6. *Lognormal mean and median WTP estimates* [(a)]

	Estimate	*Standard Error*
Location (β)	4.6221	0.08856
Scale (σ)	0.9997	0.08081
Log-Likelihood	-205.430	
	Point estimate	90% Confidence interval estimate
Mean	76.2 €	[58.2 € – 101.5 €]
Median	46.2 €	[39.9 € – 53.5 €]

Notes: [(a)] Calculations are performed using the PROC LIFEREG procedure in *SAS*®.

---

[27] CAMERON, QUIGGIN (1994).

A large number of possible predictors are available for integration in the WTP function. Formal testing procedures, based on the used of the log-likelihood test statistic, show that the model specification as presented in Table 7 fits the data the best. The estimation results show that respondents who visit the *Zandvoort* beaches two to three times per year – which corresponds to the second lowest visiting frequency presented – and plan to stay at the beach the whole day are willing to pay more than the average respondent. Second, all respondents who visited the beach during the weekends and choose to stay at the *Bloemendaal* beach are willing to pay more than the average respondent. By contrast, respondents who visit the site during the winter are willing to pay, on average, less for the described protection programme. This may be due to the fact that the marine biological pollution, as described in the questionnaire, is less likely in the winter than in the summer, because of the lower temperature of the water. In addition, all respondents who ranked the protection of coastal reserve areas 'the most important priority for beach management' are willing to pay more for the marine protection programme than the average respondent. Finally, respondents who spent a longer time travelling or who incurred higher parking costs have lower WTP. This suggests that the values obtained with the TC and CV methods are largely complementary. In other words, the TC exercise has captured other value categories than the CV exercise.

All in all, these results show that preferences for a marine protection programme are not independent of the respondent 'profile'. The consequence of this fact, from the policy design perspective, is that the marine protection programme is not equally desired among the Dutch population. For example, the stronger supporters of the programme are the taxpayers who live nearby and visit the beach on a regular basis (*e.g.* on the weekends). By contrast, the taxpayers who incur relatively high travel costs, *e.g.* who spent a longer time traveling or who incurred higher parking charges, are willing to pay less for the ballast water treatment programme.

Table 7. *WTP function* [(a)]

	Estimate	*Std. Error*	*p-value*
Beach attributes			
Visited area: *Bloemendaal*	0.206	0.33	0.53
Day of the visit: week-end	0.418**	0.24	0.08
Recreational profile of the visitor			
Number of adults in the group	0.093	0.09	0.30
Presence of a child in the group	0.086	0.30	0.77
2 or 3 visits per year	0.633*	0.31	0.04
Visits in the winter season	– 0.534*	0.23	0.02
Time planned to stay on the beach: all day	0.427**	0.24	0.08
Marine management policy options			
Nature reserves closed to the public	0.271**	0.16	0.09
Infrastructure support	0.358*	0.13	0.01
Socio-economic characteristics			
Age	– 0.005	0.08	0.52
Income	0.008	0.01	0.61
University degree	0.169	0.24	0.48
Education in economics or business management	– 0.288**	0.22	0.20
Travel costs and on-site expenditures			
Transport	0.003	0.00	0.55
Travel time	– 0.014**	0.01	0.20
Parking fee	– 0.056*	0.02	0.01
Expenditures on beach materials	0.021**	0.01	0.06
Expenditures on food and drinks at the beach-house	0.005**	0.00	0.18
Lognormal parameters			
Intercept	3.323		
Scale	0.870		
Log-Likelihood	–116.40		

Notes: [(a)] Calculations are performed using the PROC LIFEREG procedure in *SAS®*.
* Significant at 5%
** Significant at 10%

5.2.4. Policy implications

The assessment of the individual recreation benefits is performed by using the travel cost model methodology so as to get a standard (Marshallian) demand curve for yearly visits to the *Zandvoort* beach area. The underlying consumer surplus of recreation corresponds to area underneath the resulting demand curve. In order to assess such a consumer surplus, we evaluate all the explanatory variables of the demand function at their sample mean, with the exception of the individual travel costs variable[28]. The generalized travel costs variable is defined as the sum of transport, travel time and parking costs, which is denoted by P in the following reduced form of the demand curve[29],

$$P = 67.7153 - 41.6777 \times Log\, N$$

where P denotes the generalized travel cost and N the yearly number of trips. Integrating such an equation we get a gross recreational benefits per individual estimated to be 55 € per year. In other words, we can predict that if the beach area of *Zandvoort* is closed to visitors for an entire year the total recreational welfare loss would equal, on average, 55 € per individual.

This study has also offered an economic value assessment of the non-market benefits of a marine protection programme. This programme focuses on the prevention of harmful algal blooms. The contingent valuation estimates indicate that the annual WTP amounts to 76 € per respondent per year. The comparison of the TC and CV estimates indicate the importance of marine ecosystem non-market benefits, even if we admit the presence of double counting in the CV estimates. Furthermore, the combination of TC and CV estimation results shows that if no policy action is undertaken to prevent a HABs marine pollution event in the coast of the Netherlands a significant welfare loss may result. An estimate of the total welfare loss amounts to 326,190,000 €, which corresponds to 0.08 percent of the Dutch GDP measured at market prices for the year 2000. This value is obtained by: (i) assuming that the respondents that participated in the survey are representative of the entire visitor population of the beaches along the North-Holland coastline; and (ii) multiplying the sum of the use recreational benefits, which is derived from the TC model and amounts to 55 €, by the marine ecosystem benefits, which is assessed with the CV exercise and amount to 76 €, by the total number of visitors to the North-Holland coastline, estimated at 2,400,000 per year (CBS 2000). In addition, if we assume that the respondents who refused to participate in the survey have a zero WTP, then the total welfare loss amounts to 225,071,100 €. The two estimates can be interpreted as upper and lower bounds to the financial cost of implementation of an efficient marine protection programme, respectively. In other words, the marine protection programme can be defended from a cost-benefit perspective if its cost is in the range 225 - 326 million euro. For example, if we assume that the costs of cleaning sea water and coastal beaches have a similar magnitude to the clear up programme of polluted water soils in the Netherlands, the annual costs of which amount to 0.03 percent over the period 1994-1998 (CBS 1999), then we can conclude that the benefits from such a marine protection programme far exceeds the costs and is thus recommended from a cost benefit perspective.

[28] See LOOMIS and WALSH (1997) for further discussion.

[29] See for additional econometric details, NUNES, P.A.L.D. and VAN DEN BERGH, J.C.J.M (2002).

5.3. Study 2: Cockle fishery in the Wadden Sea; a stated choice survey

The Wadden Sea is a natural area that stretches out along the north coasts of the Netherlands and Germany, and the west coasts of Denmark. Since 1993 the primary function of the Wadden Sea is as a natural area[30]. This implies that human activities are allowed as long as they do not cause significant harm to natural values of the Wadden Sea. However, shell fishing has always been present in Dutch coastal area, and in the Wadden Sea in particular. In the early 20th century, coinciding with the mechanization of fishing effort, the harvest started to increase to such an extent that it exceeded local consumption. This stimulated export of shellfish, which in turn has provided a stimulus to fishing efforts, ultimately leading to the introduction of mechanical equipment, such as the hydraulic trawl. According to environmental organizations (*e.g.* De Waddenvereniging[31]), the process of mechanical shell fishing altered the sediment structure of the seabed irreversibly[32, 33]. An additional negative impact of cockle fishery is the withdrawal of a great amount of cockles from the food web in the Wadden Sea. In particular, cockles constitute an important element of the diet of the bird population in the Wadden Sea[34].

In order to minimize the negative environmental impact of shell fishery, the current fishery is regulated by the central government. The overall objective is to define precise food requirements, on the advice of biological experts, for the bird species wintering in the Wadden Sea. These can be translated into harvest standards for the cockle fishery. One policy proposal on the government's agenda is identification of sensitive areas in the Wadden Sea, in which fishing is not allowed. Many individuals and organizations (including politicians and environmental groups such as de Waddenvereniging and Wilde Kokkels) propose total ban on cockle fishing in the Wadden Sea. Nevertheless, the perception and evaluation of the impacts of the cockle fishery is not the same for all stakeholder groups. In fact, current public debate about cockle fishery in the Wadden Sea makes it clear that opinions differ about whether current cockle fishery is ecologically sustainable.
So the current fishery policy involves two main restrictions on fixed areas where it is (not) allowed to fish; and quota to save food for birds. There is no consensus about whether these measures are sufficient to realize the ecological sustainability of the Wadden Sea ecosystem. For this study we propose additional, hypothetical but feasible, fishery management options for regulation of the cockle fishery in the Wadden Sea. We choose to include the following three groups of policy measures: area policy measures, *i.e.* the surface area where it is allowed to fish; quota policy measures, *i.e.* the maximum amount of cockles the cockle-fishery sector is allowed to fish; and, rotation policy measures, the exclusion of areas from

---

30 LNV: Ministerie van Landbouw, 1993.
31 Source: http://www.waddenvereniging.nl.
32 Source: http://www.wildekokkels.nl.
33 STICHTING ODUS (2001), *Uit de Schulp: Visie op duurzame ontwikkeling van de Nederlandse Schelpdiervisserij*, Stichting Odus, Yerseke.
34 EVA II (2003), *Resultaten wetenschappelijk onderzoek EVAII; publieksversie*, Ministerie van Landbouw, Natuur en Voedselkwaliteit, Den Haag.

the fishing activities on a rotation basis. This last policy option, *rotation*, has been proposed by the cockle fishermen. They see it as an opportunity to increase the harvest as well as improve sustainability of fishing[35]. Natural scientists, however, have suggested that rotation implies a disturbance of the ecosystem that is more widely spread in the Wadden Sea. The environmental benefits of all policy measures considered are an increase in the number of birds. This increase is regarded as an indication of the quality of the ecosystem. A relevant question is whether different stakeholders prefer similar levels of ecosystem quality or, alternatively, each stakeholder has her own preferences about the ecosystem quality of the Wadden Sea area.

5.3.1. Survey instrument and survey execution

In each questionnaire the three different policy measures for regulation of cockle fishery are presented to the respondent. These policy measures refer to the three different variables: 'area', 'quota', and 'rotation'. A fourth variable gives the changes in the number of birds, which is interpreted as a proxy for ecosystem quality. The last variable in the choice experiment is a monetary variable and refers to a one time lump-sum amount for an associated policy option. This gives the total of five distinct variables for our stated choice valuation exercise, as presented in Table 8. To set the level of the attributes for the policy attributes and the level of birds attribute, advice by a marine biologist with expertise in the study area was used.

Furthermore, in this study, the preferences of the individuals belonging to the different stakeholder groups will be examined in systematically. In other words, the question who will benefit and lose from specific policies is addressed by measuring the welfare gains and welfare losses associated with them. Stakeholder groups include cockle fishermen, Dutch citizens, local residents, tourists, policy makers and natural scientists. Local residents are the people living in the coastal area of the Wadden Sea: inhabitants of the Dutch Wadden islands, and people who live less than five kilometers from the coast in the northern provinces of the Netherlands. The policy makers are local politicians, local and national civil servants and members of lobby groups involved in preparing formal policy[36]. Natural scientists include biologists at universities or other (scientific) research institutes who have a special interest in the Wadden Sea.

The combination of all stated choice attributes and respective attribute levels allowed us to create a full factorial design. The outcome of such a procedure combined with the elimination of all policy combinations that turn out to be internally inconsistent as well as the results from the a series of pre-tests resulted in 168

[35] STICHTING ODUS (2001), *o.c.*

[36] This is an example of the 'polder model' (a consulting model), that is used for public decision making in the Netherlands.

different policy combinations. Presenting each respondent with 168 choice questions is not a feasible option.

Table 8. *Attributes and their levels used in the survey*

Stated choice attributes	*Attribute's levels*
Policy measure 1: 'area'	
Surface area where it is allowed to fish for cockles	The whole Wadden Sea Current level Half of the current level (Nowhere; the banning option)
Policy measure 2: 'quota'	
Allowed number of cockles harvested.	Current level Lower level
Policy measure 3: 'rotation'	
Rotation or fixed areas where it is allowed to fish for cockles	Rotation No rotation (current situation)
Change in number of birds (indication of the quality of the ecosystem)	Lower level Current level More than in current situation Much more than in current situation.
Costs per household	9 different money amounts between zero en 250 euro

We choose to present eight different choice questions to each respondent. The 168 sets are then divided over 21 versions of eight questions, which were distributed randomly among respondents. In addition, we consider the 'extreme policy scenario' (a ninth question). This corresponds to entirely banning the mechanical cockle fishery sector from the Wadden Sea area. To make this policy scenario realistic, the survey design always portrays to the respondents a higher level of birds, indicating a better ecosystem quality. In all policy combinations presented to the respondent, a 'non-choice' option was always included. Omission of this option can yield biased and misleading WTP estimates[37]. In short, each respondent faced nine trade-offs between two alternatives, the current situation and a policy option. For the current situation, the price is always equal to zero (*i.e.* the price is normalized to the *status quo*). For the alternative situation, the price is always greater than zero. Including the

[37] LOUVIERE, J.J., & STREET, D.J. (2000), Stated-preference methods, in: HENSHER, D.A., & BUTTON, K.J. (eds.), *Handbook of Transport Modelling*, Elsevier Science.

current situation in each stated choice question allows us to identify whether stakeholders prefer to keep the status quo policy scenario.

Interviewers for the tourists and local residents were recruited among students. The tourists were interviewed in the summer of June to August 2003, while the local residents were interviewed in autumn, October to November 2003. The sample sizes for the tourists and the locals are 332 and 420, respectively. Most of the tourists were interviewed on the Islands Texel and Vlieland, and on the ferries to these islands. The local residents interviewed live in various parts of the Wadden Sea area. The local residents were interviewed on the Islands Schiermonnikoog and Terschelling and in the northern provinces Friesland, Groningen and North-Holland. The sample of policy makers was constructed by collecting relevant addresses via the Ministry of agriculture and via the Internet. To construct the sample of natural scientists, addresses were collected via dr. B. ENS, chairman of a large scientific research group responsible for a recent report on the impact of shell fishery for the Ministry, as well as via the Internet. Both policy makers and natural scientists were sent the questionnaire by mail.

Table 9 summarizes some characteristics of the respondents representing the five stakeholder groups. The characteristics of the different stakeholder groups differ. The policy makers and the natural scientists have on average a larger income than the other two groups.

Table 9: *Information about the stakeholder groups*

	Dutch citizens	*Tourists*	*Local residents*	*Policy makers*	*Natural scientists*
Sample size	1558	332	420	39	29
Income					
Low (<1500 euro)	28%	14 %	25 %	0 %	3 %
Middle	54%	39 %	46 %	33 %	31 %
High (>3000 euro)	18%	33 %	12 %	64 %	66 %
Consume fish	90%	90 %	85 %	97 %	90 %
Consume cockles	11%	4 %	6 %	21 %	48 %
Know fisherman	11%	13 %	51 %	62 %	52 %
Member of environmental association	34%	56 %	45 %	82 %	90 %
Average age (in years)	40.2	42.7	43.4	50.1	48.2

The local residents have on average the lowest income. It striking to see that the highest percentage of natural scientists consume cockles. A large part of this group had only tries them once, just to know how cockles taste. More than 50 % of the local residents know a fisherman. This is also the case for the policy makers and natural scientists, but we guess that they know them in a work relation, while the local residents know them in their private life. The average age of the policy makers and the natural scientists is higher than that of the tourists and locals.

5.3.2. Estimation results

The random utility model is estimated with a conditional logit model. We do not operationalize the use of the qualitative attribute levels, using the *'status quo'* as the omitted variable for each attribute under consideration, as dummy variables. Instead, we made use of effect codes. In this way the omitted levels of each attribute can be estimated[38]. The estimation results, across the five separate stakeholder groups, are presented in Table 10. Preferences regarding policy measures are not the same for the stakeholder groups under consideration. Because a monetary variable is included in our valuation exercise, it is possible to estimate the welfare changes due to the different policy measures. Table 11 shows the welfare changes, measured in monetary terms, associated with the potential adoption of the alternative policy measures.

The model specification allows us to capture consumer preferences with respect to the *status quo* situation. A positive estimate indicates that choosing the *status quo* situation, independently of the proposed policy alternatives, increases respondents, indirect utility. In other words, respondents prefer the current cockle fishery situation. According to Table 4, Dutch citizens, tourists, local residents and policy makers have a positive estimate for the constant term. However, such an estimate turns out to be statistically insignificant from zero for the policy makers. This estimation result can be interpreted as a signal indicating that these stakeholder groups are averse to a policy change, *i.e.* they prefer the things the way they are. By contrast, natural scientists show a clear preference for a policy change since the constant term estimate is negative. Nonetheless, this parameter estimate is not significantly different from zero.

As far as the 'area' attribute is concerned, the preferences of the five groups are comparable, for the less extreme measures. In fact, all stakeholder groups prefer the policy option characterized by cockle fishing 'half of the current area' over the 'current situation'. In addition, all stakeholder groups show consistent preferences for the 'current situation' over a fishery policy that brings along with it the possibility to fish in the 'whole area' of the Wadden Sea. In short, the ranking of stakeholder preferences with respect to cockle fishing is 'half of the current area' over the 'current situation' over the 'whole area'.

One stated choice question includes the scenario regarding a ban on all the cockle fishery activities in the Wadden Sea. This scenario is interpreted as an extreme policy option. With respect to this extreme policy measure, banning cockle fishery from the Wadden Sea, the preferences differ. Policy makers and natural scientists are in favor of banning cockle fishery, while the other groups prefer the current situation over this extreme scenario. Estimation results show that none of the

[38] See HOLMES, T.P., & ADAMOWICZ, W.L. (2003), Attribute-based Methods, in: CHAMP, P.A., BOYLE, K.L., & BROWN, T.C. (eds.), A *Primer on Nonmarket Valuation*, Kluwer Academic Publishers, Dordrecht, 171-219.

stakeholders reveal a clear preference with respect to the 'quota' policy measure. In fact, Table 10 shows that the respective parameter estimates are not statistically significant from zero.

Table 10. *Estimation results of different stakeholder groups*

	Dutch citizens	*Tourists*	*Local residents*	*Policy makers*	*Natural scientists*
Constant term	0.56***	0.24*	0.711***	0.37	-0.46
	(7.94)	(1.93)	(6.19)	(0.95)	(-1.08)
Area where is it allowed to fish					
Whole area	-0.35***	-0.28***	-0.32***	-0.30	-0.70***
	(-7.41)	(-3.35)	(-4.08)	(-1.35)	(-2.74)
Half the current area	0.20***	0.20***	0.15***	0.357**	0.603***
	(6.53)	(3.15)	(2.52)	(2.04)	(2.81)
Ban cockle fishery	-0.22***	-0.44***	-0.36***	0.24	0.82***
	(-6.77)	(-6.74)	(-5.55)	(1.25)	(2.45).
Quota					
Half the current quota	0.01	0.04	-0.05	-0.01	0.06
	(0.49)	(0.83)	(-1.30)	(-0.10)	(0.39)
Rotation of the fishing area					
Rotation	0.11***	0.10*	0.08*	-0.09	-0.29*
	(4.40)	(1.91)	(1.88)	(-0.74)	(-1.90)
Level of birds					
Fewer birds	-0.96***	-1.48***	-0.59***	-1.01	-0.41
	(-8.57)	(-6.17)	(-3.01)	(-1.62)	(-0.72)
More birds	0.68***	0.80***	0.50***	0.00	0.02
	(15.13)	(8.35)	(6.13)	(0.02)	(0.09)
Many more birds	0.70***	0.98***	0.362***	0.92***	0.59*
	(13.86)	(9.06)	(3.99)	(3.24)	(1.93)
Price	-0.01***	-0.01***	-0.01***	-0.01***	-0.01***
	(-33.17)	(-16.19)	(-11.83)	(-3.69)	(-3.76)
Log likelihood	-7493	-1712	-2147	-201	-140
No. of observations	12981	2932	3578	329	244
Adjusted ρ^2	0.17	0.15	0.13	0.09	0.14

Notes : The significance of the preference weights is indicated by ***, ** and *, referring, respectively, to the 1%, 5% and 10% level, with t-value between brackets.

We can conclude that the reduction of the current quota by half is an unattractive policy measure for all stakeholder groups. As mentioned, including rotation as a policy attribute in the management of the cockle fishery in the Wadden Sea has been originally proposed by the cockle-fishing sector. Table 10 shows that the introduction of a rotation principle is welcomed among tourists and local residents. By contrast, this policy is firmly rejected by the natural scientists. The estimation results show that the stakeholder groups clearly differ in terms of average preferences over the policy measures. While all groups prefer the policy measure to halve the area where it is allowed to fish, the tourists and the local residents do not like the 'extreme scenario' in which fishing is banned. The tourists and the local residents like the 'rotation' policy measure, whereas the policy makers, and more so the natural scientists, dislike this measure. The latter believe that this fishery policy may destroy the ecosystem in the Wadden Sea.

Table 11. *Marginal Willingness to Pay estimates (in euro)*

	Dutch citizens	*Tourists*	*Local residents*	*Policy makers*	*Natural scientists*
Constant term	41***	24*	101***	54	-62
Area where is it allowed to fish					
Whole area	-25***	-28***	-45***	-44	-94***
Current area	-26	-51	-75	97	44
Half the current area	15***	20***	21***	53***	81***
Ban cockle fishery	-16***	-44***	-51***	36	110***
Quota					
Current quota	-1	-4	8	2	-8
Half the current quota	1	4	-8	-2	8
Rotation of the fishing area					
No rotation (current situation)	-8	-10	-12	14	39
Rotation	8***	10*	12*	-14	-39*
Level of birds					
Fewer birds	-70***	-146***	-84***	-149	-55
Current level of birds	-31	-30	-38	13	-27
More birds	50***	80***	71***	0	3
Many more birds	51***	97***	51***	136***	79*

Notes : The significance of the preference weights is indicated by ***, ** and *, referring, respectively, to the 1%, 5% and 10% level, with t-value between brackets.

5.3.3. Policy analysis

Monetary value estimates assigned by the different stakeholder groups to the different policy measures were estimated. Tourists and local residents prefer the current situation regarding the area where it is allowed to fish. Policy makers and natural scientists prefer half the current area. The natural scientist prefer to ban cockle fishery in the Wadden Sea. The WTP for a smaller quota is almost zero for all stakeholder groups. Tourists and local residents prefer the 'rotation' policy measure, while natural scientists and policy makers prefer no rotation. According to the monetary valuation results (see Table 11), natural scientists show the strongest (average) magnitude with respect to the 'area' policy. In fact, the marginal willingness to pay for adoption of the 'half the current area' policy is about 81 Euro, whereas the marginal willingness to accept to 'ban cockle fishery' policy is about 110 Euro. According to the welfare estimates, tourists and local residents present a marginal WTP for the adoption of the rotation principle that ranges from 8 to 10 Euro. In contrast, a natural scientist has on average a WTP of 39 euro to prevent the introduction of rotation. A possible explanation for this response pattern can be the fact that laypeople perceive the rotation principle as positive. In contrast, expert individuals, such as natural scientists and policy makers, clearly see the disadvantages of such a procedure. These relate to a wider spread of the human activity in such a natural marine environment, creating an additional threat to many of the sensitive marine areas. Finally, all respondents prefer more birds than in the current situation. Local residents prefer the option 'more birds' above 'many more birds'. The other stakeholder groups prefer the option 'many more birds' over 'more birds' and the latter over the current situation. The economic results for the level of birds, and related to this, the quality of the ecosystem in the Wadden Sea are that all stakeholder groups have a WTP for a higher level of birds. The WTP for 'more birds' and 'many more birds' does not differ for the Dutch residents. The WTP for local residents is the most for 'more birds', while the WTP of the other three groups is the highest for 'many more birds'.

6. CONCLUSIONS

This paper conveys two principal important messages. The first message is of a methodological nature and refers to the review of the main reasons that steer economists to an interest in the field of the economic valuation of the environmental benefits. As we have seen, policy guidance constitutes an important motivation since most of the monetary value assessment of these benefits are crucial when performing a cost-benefit analysis of any environmental protection project. Furthermore, since most of environmental benefits are not market priced, we have identify and described the wide range of economic valuation tools that the economist has access to when deciding to run an economic valuation exercise. The second message is of an empirical nature and refers to presentation of two case studies. In each valuation study, we not only explain how different economic valuation methodologies, including travel cost, contingent valuation and stated choice method, can be applied to the economic valuation of marine environmental benefits but also stresses the link between the valuation estimates and policy guidance. In other words, a economic valuation exercise needs a follow-up. In particular, the valuation results always need to be contrasted with the monetary costs regarding the (specific programme of) marine environmental protection. From a cost benefit perspective, the protection is recommended if and only if the benefits from such a marine protection programme far exceeds the costs.

REFERENCES

- ARONSSON, T., JOHANSSON, P.-O. & LOFGREN, K.-G. (1997), *Welfare Measurement, Sustainability and Green' Accounting - a Growth Theoretical Approach*, Edward Elgar, Cheltenham, UK and Lyme, US.

- BILLER, D., & BARK, B. (2001), *Valuation of Biodiversity: Selected Studies*, OECD, Environment Directorate, Paris, France.

- BOARDMAN, A., GREENBERG, D., VINING, A., & WEIMER, D. (2000), *Cost Benefit Analyis: Concepts and Practice*, 2nd Edition, Prentice Hall.

- CAMERON, T.A., & JAMES, M.D. (1987), Efficient estimation methods for closed-ended contingent valuation surveys, *American Journal of Agricultural Economics*, 269-276.

- CARSON, R.T., MITCHELL, R.C. , HANEMANN, W.M., KOPP, R.J., PRESSER, S., & RUUD, P.A. (1992), *A Contingent Valuation Study of Lost Passive Use Values Resulting from the Exxon Valdez Oil Spill*, Report prepared for the Attorney General of the State of Alaska, Washington, US.

- EL SERAFY, S. (1999), *Steering the Right Compass: Proper Measurements for Sound Macro-economic Management*, paper presented at International Symposium Valuation of Nature and Environment, Royal Netherlands Academy of Arts and Sciences, Amsterdam, The Netherlands.

- EVA II (2003), *Resultaten wetenschappelijk onderzoek EVAII; publieksversie*, Ministerie van Landbouw, Natuur en Voedselkwaliteit, Den Haag.

- GERLAGH, R., DELLINK, R., HOFKES, M., & VERBRUGGEN, H. (2002), A Mesure of Sustainable Development for the Netherlands, *Ecological Economics*, 41, 157–74.

- HANEMANN, M.W., LOOMIS, J., & KANNINEN, B. (1991), Statistical Efficiency of Double-bounded Dichotomous Choice Contingent Valuation, *American Journal of Agricultural Economics*, 73(4), 1255-1263.

- HANLEY, N., & SPASH, C.L. (1993) *Cost-Benefit Analysis and the Environment*, Edward Elgar, Aldershot, UK and Brookfield, US.

- HANLEY, N., MOURATO, S. *et al.* (2001), Choice modelling approaches: A superior alternative for environmental valuation?, *Journal of Economic Surveys* 15(3), 435-462.

- HANLEY, N., SHOGREN, J., & WHITE, B. (2001), *An Introduction to Environmental Economics*, Oxford University Press, Oxford, UK.

- HOLMES, T.P., & ADAMOWICZ, W.L. (2003), Attribute-based Methods, in: CHAMP, P.A., BOYLE, K.L., & BROWN, T.C. (eds.), A *Primer on Nonmarket Valuation*, Kluwer Academic Publishers, Dordrecht, 171-219.

- HUETING, R. (1980), *New Scarcity and Economic Growth*, North-Holland, Amsterdam, The Netherlands.

- KEUNING, S.T., & DE HAAN, M. (1996), *What is a NAMEA?*, Statistics Netherlands National Accounts Paper, Voorburg, The Netherlands.

- KRUPNICK, A., ALBERINI, CROPPER, M., SIMON, N., O'BRIEN, B., GOEREE, R., & HEINTZELMAN, M. (2000), *Age, Health, and the Willingness to Pay for Mortality Risk Reductions: A Contingent Valuation Survey of Ontario Residents*, Resource for the Future Discussion Paper, n. 00–37, Washington D.C., US.

- LNV (Ministerie van Landbouw, Natuurbeheer en Visserij) (2003), *Vissen naar Evenwicht: Regeringsbeslissing Structuurnota zee-en kustvisserij*. Ministerie van Landbouw, Natuurbeheer en Visserij, Den Haag.

- LOUVIERE, J.J., & STREET, D.J. (2000),. Stated-preference methods, in: HENSHER, D.A., & BUTTON, K.J. (eds.), *Handbook of Transport Modelling*, Elsevier Science.

- MANSFIELD, C., VAN HOUTVEN, G., JOHNSON, F.R., CRAWFORD-BROWN, D., & ZACHARY, P. (2002), *Behavioral Reactions to Ozone Alerts: What Do They Tell Us About Willingness-to-Pay for Children's Health?*, Valuation of Environmental Impacts on Children's Health, EPA Research Project, Washington D.C., US.

- MITCHELL, R.C., & CARSON, R.T. (1989), *Using Surveys to Value Public Goods: The Contingent Valuation Method*, Resources for the Future, Washington DC, US.

- NIJKAMP, P., RIETVELD, P., & VOOGD, H. (1991), *Multicriteria Evaluation in Physical Planning*, North-Holland, Amsterdam, The Netherlands.

- NUNES, P.A.L.D. (2002), *The Contingent Valuation of Natural Parks: Assessing the Warm Glow Propensity Factor*, New Horizons in Environmental Economics Series, Edward Elgar, Cheltenham, UK and Northampton, MA, US.

- NUNES, P.A.L.D., & SCHOKKAERT, E. (2003), Identifying The Warm Glow Effect in Contingent Valuation, *Journal of Environmental Economics and Management*, 45, 231-45.

- NUNES, P.A.L.D., & VAN DEN BERGH, J.C.J.M. (2001), Economic Valuation of Biodiversity: Sense or Nonsense, *Ecological Economics*, 39, 203–22.

- NUNES, P.A.L.D., & VAN DEN BERGH, J.C.J.M. (2004), Can People Value Protection against Invasive Marine Species? Evidence from a Joint TC–CV Survey in the Netherlands, *Environmental and Resource Economics*, 28, 517–32.

- STICHTING ODUS (2001), *Uit de Schulp: Visie op duurzame ontwikkeling van de Nederlandse Schelpdiervisserij*, Stichting Odus, Yerseke.

- WANG, H, & WHITTINGTON, D. (2000), *Willingness to Pay for Air Quality in Sofia, Bulgaria*, World Bank Policy Research Working Paper, n. 2280, Washington D.C., US.

- WANG, H., LAPLANTE, B., WU, X., & MEISNER, C. (2004), *Estimating Willingness-to-Pay with Random Valuation Models: An Application to Lake Sevan, Armenia*, World Bank Policy Research Working Paper, n. 3367, Washington D.C., US.

CHAPTER 8

A CONTINGENT VALUATION STUDY OF AN ACCIDENTAL OIL SPILL ALONG THE BELGIAN COAST

KARL VAN BIERVLIET[*], DIRK LE ROY

ECOLAS nv

Lange Nieuwstraat 43
B-2000 Antwerp
Belgium
Tel: +32 3 233 07 03, Fax: +32 3 233 81 20
E-mail: karl.vanbiervliet@ecolas.be

and

PAULO A.L.D. NUNES

Department of Economics, University of Venice, Italy
and
Fondazione Eni Enrico Mattei (FEEM), Italy

E-mail: pnunes@unive.it

[*] *Author for correspondence*

F. Maes (ed.), Marine Resource Damage Assessment, Liability and Compensation for Environmental Damage, 165–207.

1. INTRODUCTION

The North Sea is situated between Great Britain, Norway, Sweden, Denmark, Germany, The Netherlands, Belgium and France. The Belgian North Sea coast has a total length of 65 km. This coastline plays a significant role in both regional and national economies since important industrial and service activities including fisheries, sand extraction, and tourism are located in this area. On the other hand, the Belgian sea and coast is also a unique ecological system. Such an ecological system not only provides different environmental benefits, such as the protection of valuable terrestrial and marine habitats, including sandbanks, beach, flat and salt marsh, estuary, and dune biotopes, which host a wide range of mammal and plant life diversity. For instance, vulnerable sea and coastal birds have their breeding ground and wintering place on the Belgian coast. In addition, such an ecological system is characterized by a dynamic and resilient function that, in turn, provides an important ecological services such as the chemical balance of the water.
Since most of this range of environmental benefits have a price and all the industrial and service activities are market priced, human interventions have been invested in promoting these same economic activities leaving the marine and coastal ecosystem under heavier pressure.

Accidental oil spills are one example of the human pressure and constitute one of the biggest threats to that ecologic system. Annually 10,000 ships sail through the English Channel heading to or coming from the ports around the North Sea, some of them with an international character, such as Rotterdam in the Netherlands and Antwerp in Belgium. Due to the shallow depth of the Southern North Sea the large ships (oil tankers among them) are channelled in a central, deep shipping route. These shipping conditions, climatic conditions, such as dense fog, and possible human errors, cause a real risk on accidents.
In the period 1991-1998 eight shipping accidents occurred in the Belgian part of the North with oil spillage as a result. The largest accident, until now, in 1992 (*'Amer Fuji'/'Meritas'* collision) resulted in an estimated 225m^3 oil spill. Accidents with major oil pollution happen regularly in European waters. At the end of 1999, in France, the accident with the *Erika* occurred in coastal waters near Bretcony. At the end of 2002, in Spain, there was the accident with the *Prestige*. Both accidents caused enormous damage to the marine environment. For this reason the economic valuation of the damage caused is an important instrument for environmental damage assessment (*e.g.* assessing who is affected by the spill and how important the damage is in monetary terms) and policy guidance (*e.g.* assessing which type of government intervention to implement – should one invest in prevention or restoration activities?).

Following the general framework of environmental economics, we can distinguish between damages to use and non-use functions. In case of an oil spill, the

most important use functions are commercial fishing and recreation activities. The data for evaluation of these functions can be respectively based on market data and alternative valuation methods such as the travel cost method[1].
In addition to use values, a marine and coastal ecosystem has an important non-use, existence or passive use value, which may be affected by an accidental oil spill. Existence values are defined as the benefit received from simply knowing the resource exists, even if no use is made of it. If the marine environment is affected by the disaster, *e.g.* the extinction of locally protected seabirds, our welfare is negatively affected. This welfare change can be estimated in monetary terms by using the Contingent Valuation (CV) method. In a CV study, a contingent valuation questionnaire is designed[2]. If a respondent was directly asked '*How much would you be prepared to pay for the non-operating function of the Belgian part of the North Sea?*', it would be very difficult for that respondent to know what they would be paying for. That is why the non-use function of the Belgian part of the North Sea is 'translated' into something more tangible; something the respondent is more familiar with. For this reason, economists work together with natural scientists to create an oil protection scenario that is described in the instrument survey and well understood by the respondents. This method has among other things been used for the estimation of non-use values in case of the *Exxon Valdez* oil spill[3]. In the study present for discussion in this chapter, the CV method has been utilized in case of an oil spill before the Belgian Coast.

2. SURVEY DESIGN AND FINAL STRUCTURE OF THE QUESTIONNAIRE

The CV method is subject to lots of criticism. The question is whether the CV method can generate valid economic values. The CV method was critically analyzed in 1993 by a committee, appointed by the National Oceanic and Atmospheric Administration (NOAA) and chaired by two Nobel Prize winners. The committee concluded that the CV method can generate valid economic values, but it is necessary that the structure of the survey fulfil certain requirement[4]. During the development of this CV study, the NOAA criteria were taken into consideration.

The development of the CV survey instrument took over 10 months. A first survey was formulated by means of a literature study. Because of the need for a reliable

---

1 LOOMIS, J.B & WALSH, R.G. (1997), *Recreation Economic Decisions: Comparing Benefits and Costs*, Pennsylvania, USA.
2 See NUNES, P.A.L.D. (2002), *The contingent valuation of natural parks: assessing the warmglow propensity factor, new horizons in environmental economic series*, Edward Elgar Publishing, UK
3 CARSON, R.T., MITCHELL, R.C., HANEMANN, M.W., KOPP, R.J., PRESSER, S. & RUUD, P. A. (1992), *A contingent valuation study of lost passive use values resulting from the Exxon Valdez oil spill. A report to the Attorney General of the State of Alaska*, 127 p.
4 ARROW, K., SOLOW, R., PORTNEY, P.R., LEANER E.E., RADNER, R. & SCHUMAN H. (1993), *Report of the NOAA Panel on Contingent Valuation*, Federal Register 58, 4.601-4.614.

scenario structure, a team of experts was brought together several times. The result was a number of background documents serving as a scientific basis for the different scenarios of the survey. As a result of these background documents, a second survey was formulated, in which the structure of the scenario was modified. Afterwards, a working visit was paid to an expert in the CV method, Professor J. LOOMIS from Colorado State University. The second survey was presented to him. With the suggestions he made, the second survey was modified. Finally, two focus groups were brought together to further amend the survey. In the next step, eight in-depth interviews were executed. Knowing the results of these interviews, a final survey was formulated. Minor adjustments were made to the final survey after the pre-tests. Below the final structure of the questionnaire is described with special attention to the design of the scenario and design of the valuation questions. Annex 1 presents an example of a questionnaire.

2.1. Initial questions

The interview starts with a question which checks the social attitude of the respondent with regard to a number of social problems. At the start of the interview the respondent does not know that the specific subject of the interview concerns oil pollution. The second question builds further on the first question. Here the respondent must choose between a number of specific environmental topics, including oil pollution at sea. These first two questions also serve to introduce questions. They help the respondent to empathize and to put him at his ease.
After answering both questions, the respondent is conducted slowly to the core of the interview, namely the scenario and the program with the willingness-to-pay (WTP) questions. Mainly the respondents get an explanation about the use of the opinion of the citizen for the government. At the end he is informed about the real contents of the interview: a programme for prevention of environmental damage caused by accidental oil pollution in front of the Belgian Coast.

2.2. Presentation of the reference situation and the scenario

2.2.1. Presentation of the Belgian part of the North Sea
First of all the area in the North Sea is indicated for which Belgium is responsible. Then the different functions of the North Sea, the economic, the recreational and the ecological function of the North Sea are explained with a number of examples. Next, the most important parts of the North Sea are discussed with their unique function for nature, among other things the sand banks, the beaches and the nature reserves the Zwin and the IJzermonding. Everything is shown on drawings or photographs, in a way that allows the respondent to process the information more easily.
A number of threats to this natural wealth are also listed: overfishing, pollution with environmentally dangerous substances, disruption of nature areas and oil pollution by accidents with ships and tankers. The respondent is informed that a number of

accidents with oil pollution have already happened off the Belgian coast, but with limited damage compared with damage by accidents in surrounding countries.
Then, as a consequence, it is stated that within a number of years there will almost certainly be an accident with damage to the marine environment of the Belgian part of the North Sea, caused by the narrow channels, frequently occurring fog, bad weather conditions and possible human errors.
Hereafter it is stated that international legislation will require a number of measures to prevent oil damage as from 2010. Because of this the chance of oil pollution with severe impact on the surrounding environment will be very low as from 2010. To avoid an accident in the meantime, the Belgian government can put a prevention action and intervention programme into effect. In that way the nature value of the Belgian part of the North Sea is protected as far as possible.

2.2.2. Financing the programme

To be able to finance the Belgian programme, a financial contribution is expected from both the producer and the consumer of oil products. It is also shown that almost every Belgian citizen is an oil user: products such as fuel for heating and transport, medicines, plastics and shampoo. The oil companies would pay the functioning costs of the intervention programme and the Belgian citizens would pay the investment cost of the intervention and prevention programme.
It is explained that each family would have to pay a one-time financial contribution and this approximately four months after the completion of the interview. This means that if the interview is conducted in April, the respondent should pay the financial contribution in September.
It is then stated that it is not certain that the programme will be implemented, but that it depends on a referendum. Afterwards, it is clearly explained that the aim of this interview is to examine if the Belgian population would vote for or against the intervention and prevention programme in such a referendum.

2.2.3. Presentation of the intervention and prevention programme

In this part of the questionnaire the scenario is presented with clear photographs and illustrations. The scenario in this questionnaire consists of a risk at an accidental oil spill off the Belgian Coast, with a certain environmental impact, which can be avoided with an intervention and prevention programme.
The impact is presented by means of five parameters:

- (i) number of birds that will die;
- (ii) number of fish, crabs, shrimps and lobsters that will die;
- (iii) pollution of the beach;
- (iv) pollution of the nature reserve the Zwin;
- (v) pollution of the nature reserve the IJzermonding.

A range of measures is necessary to prevent damage to the marine environment. These measures are split up in two parts. First, a prevention programme which must prevent accidents, consisting of (i) a separate shipping route and (ii) a communication system. Secondly, an intervention programme is presented which must minimize damage resulting from a possible accident. This programme consists of a set of four measures: (i) number of tugboats; (ii) number of oil-combating platforms; (iii) measure to close the nature reserve the IJzermonding; (iv) measure to close the nature reserve the Zwin.

The survey scenario is always a change in environmental quality opposed to the reference situation. This means that a CV study values a certain change in quality of the good. To check how the WTP changes in function of the environmental quality, different scenario's need to be developed.
The three scenarios that were elaborated are described in Table 1.

Table 1. *Description of the different scenarios.*

Features	*Severe scenario*	*Moderate scenario*	*Light scenario*
	Damage		
Size of the oil spill	10 000 m^3	5 000 m^3	200 m^3
# of birds that will die	43 000 – approximately 65 %	20 000 - approximately 30 %	3 500 – 5 %
# of fish, crabs, shrimps and lobsters that will die	20%	10%	0%
Pollution of the beach	60 km – 90 %	25 km – 40 %	0 km – 0 %
Pollution of the nature reserve the Zwin	Yes	No	No
Pollution of the nature reserve the IJzermonding	Yes	Yes	No
	Programme		
Separate shipping route	20 km	10 km	5 km
Communication system	Yes	Yes	Yes
# of tugboats	3	2	1
# of oil-combating platforms	3	2	1
Measure to close the nature reserve the IJzermonding	Yes	No	No
Measure to close the nature reserve the Zwin	Yes	Yes	No

In order to have a sufficient large sample per scenario, more than one scenario was incorporated in most of the questionnaires. Each respondent got two different scenarios in the survey (in some cases only one: version 4). A variation in the frequency of appearance of an accidental oil spill was also included in the questionnaire to check whether this would have an effect on the willingness to pay. The frequency of possible accidents differs from questionnaire to questionnaire; an accident will occur every three, five or ten years. The four different versions of the questionnaire are shown in Table 2.

Table 2. *Overview of the different questionnaire versions.*

Questionnaire version	*Scenario 1*	*Scenario 2*	*Frequency*
Version 1	Light (A)	Moderate (B)	1 in 3 years
Version 2	Light (C)	Severe (D)	1 in 5 years
Version 3	Moderate (E)	Severe (F)	1 in 10 years
Version 4	Severe (G)	-	1 in 10 years

2.3. Valuation questions

2.3.1. Willingness-to-pay questions

The design of the WTP questions was almost similar to the design of two earlier CV oil spill studies in the US[5].

The WTP questions are asked in the form of a referendum. The contribution only needs to be paid when more than 50 % of the Belgian households agree with the programme. A one-time-contribution is required. The payment vehicle is a payment to a fund. The money in this fund can only be used for the execution of the programme.

The 'Dichotomous choice with one follow-up' was chosen as the most appropriate elicitation method and this for two major reasons. Firstly, respondents have difficulties to put a value on the program without any assistance[6]. Secondly, the statistical power of the estimation of the willingness to pay raises with a follow-up

---

[5] CARSON, R.T., MITCHELL, R.C., HANEMANN, M.W., KOPP, R.J., PRESSER, S. & RUUD, P. A. (1992), *o.c.*; CARSON, R.T., CONAWAY, M.W., HANEMANN, M.W., KROSNICK, J.A., MARTIN, K.M., MCCUBBIN, D.R., MITCHELL, R.C. & PRESSER, S. (1996), *The value of preventing oil spill injuries to natural resources along California's Central Coast*, San Diego, Natural Resource Damage Assessment Inc.

[6] MITCHELL, R.C. & CARSON, R.T. (1989), *Using surveys to value public goods: the contingent valuation method,* Washington, D.C., Resources for the Future, 97.

question[7]. Each respondent is asked whether they are prepared to pay a certain amount (starting bid) for a suggested programme to avoid damage. If the respondent answers yes, the same question is asked with a higher amount (higher follow-up bid). If the respondent answers 'no', the same question is asked with a lower amount (lower follow-up bid).
In that way we get four possible intervals for every bid card which reflects the willingness to pay of the respondent (See Figure 1). When a respondent answers 'yes-yes', the WTP is between zero and the lower follow-up bid; when he answers 'no-yes', the WTP is between the lower follow-up bid and the start bid; when he answers 'yes-no', the WTP is between the start bid and the higher follow-up bid; when he answers 'yes-yes', the WTP is higher than the higher follow-up bid.

***Figure 1**. Dichotomous choice with one follow-up*

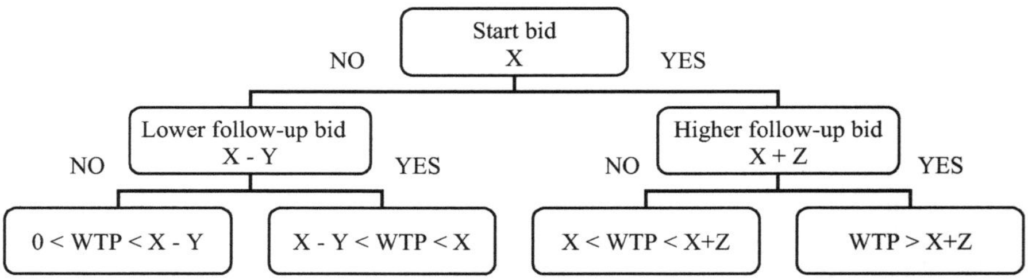

A WTP question is asked after the first and after the second scenario in every questionnaire. Seven different bid cards were used for the first scenario to gain an equal spread of amounts; the amounts on the bid cards for the second scenario were each time 20 % higher than the amounts on the bid card of the first scenario (see Table 3).

[7] CARSON, R.T., MITCHELL, R.C., HANEMANN, M.W., KOPP, R.J., PRESSER, S. & RUUD, P. A. (1992), *o.c.*, 17.

Table 3. *Different bid cards for scenario 1 and 2.*

	Bid card scenario 1			*Bid card scenario 2*		
	Start bid	*Higher bid*	*Lower bid*	*Start bid*	*Higher bid*	*Lower bid*
Bid card 1	24.79	61.97	9.92	29.75	74.37	12.39
Bid card 2	37.18	74.37	18.59	44.62	89.24	22.31
Bid card 3	49.58	99.16	24.79	59.49	118.99	29.75
Bid card 4	61.97	123.95	30.99	74.37	148.74	37.18
Bid card 5	74.37	148.74	37.18	89.24	178.48	44.62
Bid card 6	99.16	198.31	49.58	118.99	237.98	59.49
Bid card 7	123.95	247.89	61.97	148.74	297.47	74.37

2.3.2. Second round of the WTP questions

After the WTP questions the respondent gets the chance to revise his answer. This can be necessary if the respondent may have misunderstood something. For this reason it is examined if it was clear to the respondent that he would only have to pay a one-time contribution and if it was clear that he would only have to pay for one of the programmes.

If the respondent understood everything well, he also got the possibility to revise his WTP answer for the first scenario, because the respondent could change his judgment for the first scenario after being informed about the second scenario (extra information). If the respondent wished to change his WTP for the first scenario, he also got the chance to change his WTP for the second scenario.

2.4. Other questions

The last part of the questionnaire contains questions about the respondents' attitude and (possible) use of the good, evaluation questions for the respondent, questions about the respondents' socio-demographic situation and evaluation questions for the interviewer. For details about these questions we refer to the example questionnaire (annex 1).

3. SURVEY EXECUTION

3.1. Sample drawing

The relevant population for this CV study consists of all Belgian households, because the one-time contribution needs to be paid per household. The interviews

are as far as possible directed to the person who is responsible for expenditure in the household.
The sample was built up to be as far as possible representative at federal, regional and province level. The city selection in every province took into account the type of city following the 1997 hierarchy of E. VAN HECKE[8]. In every province one type of city is randomly selected. In every selected city the 'random route sample drawing with repetition' is applied. A number of start addresses for every selected city is chosen randomly. Beginning with the start address, six doorbells are rung. If nobody is at home, the interviewer goes back once on another day of the week and another hour of the day. The number of start addresses per city is determined by multiplying the size of the sample by the relative number of inhabitants per type of city and province.

3.2. Training and supervision of the interviewers

Most of the interviewers were students involved in the project through their thesis. A manual was prepared for the interviewers. The manual contained a part dealing with the CV method, an organizational part and a part with a questionnaire with an extra explanation per question. Besides this manual all interviewers got an education ranging from a half to one day. During this education some test interviews were done. Before starting with the real interviews, every interviewer had to do an internal test interview. All interviewers needed to report regularly to the coordinator of the interviews.

3.3. Survey execution

The interviews were executed in the period March-August 2001 in the Flemish Region, in the period July-September 2001 in the Walloon Region and in August 2001 in the Brussels Region. The objective was to have 500 to 600 executed interviews. In total 2,626 doorbells were rung and 571 interviews were executed (see Table 4).

In total 1,790 households were reached and 571 interviews were executed, which means an overall response rate of 32 %.

The response rate in the Flemish Region (33 %) is 2 % higher than the response rate in the Walloon Region. The low response rate in the Brussels Region can be compared with the response rate in other big cities such as Antwerp and Liège.

[8] VAN HECKE, E. (1998), Actualisering van de stedelijke hiërarchie in België, *Tijdschrift van het Gemeentekrediet*, 52ste jaar, nr. 205, Brussel, Groep Dexia.

Table 4. Response rate per district.

	Belgium		*Flemish Region*		*Walloon Region*		*Brussels Region t*	
# of doorbells rung	2 626	100 %	1 469	100 %	954	100 %	203	100 %
# of households reached	1 790	68 %	1 083	74 %	629	66 %	78	38 %
# of interviews executed	571	32 %	358	33 %	196	31 %	17	22 %

The main reasons why people do not want to cooperate are 'no interest' and 'no time'. The different interviews are divided over the different questionnaires as follows:

(i) 30 % for the questionnaire version Light-Moderate;
(ii) 30 % for the questionnaire version Moderate-Heavy;
(iii) 30 % for the questionnaire version Light-Heavy;
(iv) 10 % for the questionnaire version Heavy;

The different bid cards are divided proportionally among the survey respondents. The average duration of an interview is 26 minutes. In the Flemish Region the average is seven minutes less than in the Walloon Region and the Brussels Region. The longest interview lasted 1 hour and 20 minutes; the shortest interview lasted 10 minutes. 80 % of the interviews lasted between 20 and 40 minutes.

4. RESULTS

4.1. Descriptive analysis of the valuation questions

4.1.1. Type of response per questionnaire version

Table 6 and Table 7 present the percentage Yes-Yes (YY), Yes-No (YN), No-Yes (NY) and No-No (NN) answers per scenario and per bid card on the WTP questions. The percentage yes-yes answers on the lowest bid card is always higher than the percentage yes-yes answers on the highest bid card. This confirms the theory stating that the number of votes drop when taxes rise. The percentages of the bid cards in between do not always follow this theory consistently. The reason for that is that the different bid cards are relatively close to each other with a relatively low number of observations per bid card.

Table 6. *Answers to the WTP questions of the first scenario.*

Bid card	*Light 1/3 (A)*				*Light 1/5 (C)*				*Moderate 1/10 (E)*				*Severe 1/10 (G)*				*All 1st scenario's*			
Start bid (EUR)	*YY*	*YN*	*NY*	*NN*	*YY*	*YN*	*NY*	*NN*	*YY*	*YN*	*NY*	*NN*	*YY*	*YN*	*NY*	*NN*	*YY*	*YN*	*NY*	*NN*
24.79	60%	12%	4%	24%	46%	21%	13%	21%	55%	23%	0%	23%	30%	70%	0%	0%	51%	25%	5%	20%
37.18	35%	35%	8%	23%	28%	48%	12%	12%	42%	38%	0%	21%	45%	18%	18%	18%	36%	37%	8%	19%
49.58	28%	44%	8%	20%	41%	37%	4%	19%	36%	40%	4%	20%	50%	38%	0%	13%	36%	40%	5%	19%
61.97	37%	44%	0%	19%	13%	57%	9%	22%	25%	46%	13%	17%	36%	36%	9%	18%	27%	47%	7%	19%
74.37	20%	40%	20%	20%	29%	25%	8%	38%	27%	32%	23%	18%	33%	33%	11%	22%	26%	33%	16%	25%
99.16	23%	42%	15%	19%	29%	33%	10%	29%	22%	48%	13%	17%	11%	33%	22%	33%	23%	41%	14%	23%
123.95	13%	17%	13%	57%	23%	36%	14%	27%	23%	32%	9%	36%	13%	63%	25%	0%	19%	32%	13%	36%

Table 7. *Answers to the WTP questions of the second scenario.*

Bid card	*Moderate 1/3 (B)*				*Severe 1/5 (D)*				*Severe 1/10 (F)*				*All 2nd scenario's*			
Start bid (EUR)	*YY*	*YN*	*NY*	*NN*	*YY*	*YN*	*NY*	*NN*	*YY*	*YN*	*NY*	*NN*	*YY*	*YN*	*NY*	*NN*
29.75	56%	20%	4%	20%	38%	29%	8%	25%	41%	41%	0%	18%	45%	30%	4%	21%
44.62	38%	27%	4%	31%	28%	48%	12%	12%	33%	33%	8%	25%	33%	36%	8%	23%
59.49	32%	40%	8%	20%	44%	33%	4%	19%	36%	40%	4%	20%	38%	38%	5%	19%
74.37	26%	52%	4%	19%	9%	52%	9%	30%	24%	40%	8%	28%	20%	48%	7%	25%
89.24	12%	52%	12%	24%	25%	33%	4%	38%	19%	43%	19%	19%	19%	43%	11%	27%
118.99	27%	35%	15%	23%	33%	24%	14%	29%	30%	39%	13%	17%	30%	33%	14%	23%
148.74	13%	13%	13%	61%	27%	36%	14%	23%	18%	36%	0%	45%	19%	28%	9%	43%

4.1.2. Protest answers

A number of respondents neither want to pay the start amount nor the lower follow-up bid (No-No answer). For the first scenario in the questionnaire there are in total 129 No-No answers (= 23 %), for the second scenario there are in total 130 No-No answers (= 26 %). Respondents who give such an answer are asked for their reason. The two important reasons to give a No-No answer are 'the oil companies have to pay everything' and 'I don't want to pay extra, the government needs to pay everything'. A large number of reasons can be considered as a protest answer. This means that from the reason to answer No-No to the WTP questions, we can conclude that the respondent possible has a positive valuation for the good. The two reasons that were not considered as a protest answer are 'The proposed project has not such a big value for me' and 'My income doesn't allow me to pay that amount'. For the first scenario in the questionnaire there are in total 114 protest answers (= 20 %), for the second scenario there are in total 113 No-No answers (= 22 %) (see Table 8). The higher number of protest answers in the second scenario can be explained by the fact that some respondents declared 'the first programme is enough'. From Table 8 we can conclude the number of protest answers is higher in the Walloon and the Brussels Region. An important reason for this difference is that respondents in the Walloon and the Brussels Region state that the Flemish Region (where the Coast is situated) should pay everything. No conclusions can be drawn about the very high rate of protest answers in the Brussels Region since the number of observations is very low.

Table 8. *Protest answers per scenario and per district.*

Protest answers	*Belgium*		*Flemish Region t*		*Walloon Region t*		*Brussels Region*	
	#	%	#	%	#	%	#	%
1st scenario	114	20 %	64	18 %	42	21 %	8	47 %
2nd scenario	113	22 %	62	20 %	42	24 %	9	69 %

4.1.3. Second round of the WTP answers

Most of the respondents (98 %) understood well that they had to pay a one-time contribution for one of the two presented projects. 5 % of the all the respondents revized their answer to the WTP in the first scenario, for the second scenario this was less than 1 %. Only 15 % of the respondents who stated that they did not understand that it was a one-time payment revised their answer. For respondents stating that they did not understand that they only had pay for one scenario, the revision percentage was 25 %. Two thirds of the respondents revising their WTP for the first scenario, changed their WTP answer to a No-No answer with as main reason that they preferred the second scenario above the first scenario.

4.2. Calculation and analysis of the revealed WTP

4.2.1. Methods for calculation of the revealed WTP

The revealed willingness to pay is estimated in a parametric way and a non-parametric way.

No assumption is made about the form of the underlying distribution when the revealed WTP is calculated in a non-parametric way. The 'Turnbull likelihood estimation approach'[9] is used for the estimation of the cumulated density function of the WTP in the intervals defined by the start bids and the higher and lower follow-up bids of the different bid cards[10].

A logit regression model is used for the parametric estimation of the revealed WTP. Such a model is able to work with a discrete dependent variable, in contrast to a linear regression model. We used the software developed by COOPER and HELLERSTEIN (US Department of Agriculture, Economic Research Service) for the analysis of 'double-bounded dichotomous choice CVM-studies'.[11] This program uses the 'maximum likelihood estimation with the analytic first and second derivatives' of HANEMANN *et al.*[12] for the estimation of 'double-bounded logit' coefficients.

4.2.2. Calculation of the average and total WTP

The average WTP per household is calculated in two ways:

(i) Non-parametric (Turnbull likelihood estimation approach). The 'Lower-bound mean' of the estimated cumulative density function is calculated (conservative average). When a cumulative density function looks as follows:

WTP-intervals (EUR)	*Cumulative density percentage*
60 - ∞	100 %
40 - 60	90 %
20 – 40	30 %
0 - 20	10%

[9] TURNBULL, B.W. (1976), The empirical distribution function with arbitrarily grouped, censored and truncated data, *Journal of the Royal Statistical Society*, B38, 290-295.

[10] These calculations were executed with the Turnbull nonparametric density estimation for CVM - Gauss version 1.0, October 1996, Olvar Bergland - Commented and extended by Paulo NUNES (1997).

[11] Referendum CVM Programs, June 1994, DBLOGIT.

[12] HANEMANN, M.W., LOOMIS, J. & KANNINEN, B. (1991), Statistical efficiency of double-bounded dichotomous choice contingent valuation, *American Journal of Agricultural Economics*, 73(4), 1.255-1.263.

Then the 'Lower-bound mean' is calculated using the following formula:
0 * 0.1 + 20 * (0.3 – 0.1) + 40 * (0.9 – 0.3) + 60 * (1 – 0.9) = 34 euro.

(ii) Univariate–parametric (logit-model), the average WTP is calculated out of the coefficients of the generated logit-function:

$$\frac{\text{- Coefficient constant}}{\text{Coefficient bid amount}}$$

The average WTP is calculated for every scenario with the protest answers included and excluded.
From Table 9 we can conclude that the average WTP (protest answers included) calculated with the logit-model and the Turnbull likelihood estimation approach are close to each other (maximum difference = 6.96 euro in the Heavy-1/5 scenario).
De average WTP (protest answers included) varies from 88 euro per household in the Light-1/3 scenario to 112 euro per household in the Heavy-1/5 scenario (difference of 23.7 euro). The differences between the different scenarios are small (see also paragraph 4.2.3.).

Table 9. *WTP per household in function of the scenario's, protest answers included (in EUR).*

Scenarios	*Light*		*Moderate*		*Severe*	
	Parametric - Logit	*Non-parametric*	*Parametric - Logit*	*Non-parametric*	*Parametric - Logit*	*Non-parametric*
1 in 3 years	88.37	90.31	100.00	103.10		
1 in 5 years	89.22	93.33			105.11	112.07
1 in 10 years			97.67	97.1	106.97	110.21
					98.02	94.50

The same can be concluded for the sample without protest answers (see Table 10).

***Table 10**. WTP per household in function of the scenario's, protest answers excluded (in EUR).*

<table>
<tr><th rowspan="2">Scenarios</th><th colspan="2">Light</th><th colspan="2">Moderate</th><th colspan="2">Severe</th></tr>
<tr><th>Parametric - Logit</th><th>Non-parametric</th><th>Parametric - Logit</th><th>Non-parametric</th><th>Parametric - Logit</th><th>Non-parametric</th></tr>
<tr><td>1 in 3 years</td><td>116.16</td><td>115.79</td><td>134.83</td><td>133.84</td><td></td><td></td></tr>
<tr><td>1 in 5 years</td><td>117.13</td><td>116.81</td><td></td><td></td><td>142.24</td><td>142.86</td></tr>
<tr><td rowspan="2">1 in 10 years</td><td rowspan="2"></td><td rowspan="2"></td><td rowspan="2">119.19</td><td rowspan="2">117.13</td><td>137.43</td><td>135.99</td></tr>
<tr><td>112.59</td><td>108.87</td></tr>
</table>

When the averages are converted to the Belgian population (average per family multiplied with the number of families), the total one-time willingness to pay of the Belgian population varies, protest-answers excluded, between 492 million euro and 606 million euro and protest-answers included, between 375 million euro and 476 million euro. If you assume that the people who refuse to participate in the questionnaire have a zero WTP, then these values vary between 157 million euro and 194 million euro and protest-answers included, between 120 million euro and 152 million euro.

4.2.3. Comparative analysis

To verify whether the differences in willingness to pay are statistically robust to alternative survey design specifications, we performed formal testing procedures regarding the sensitivity of the WTP regarding:

(i) order of the oil spill, and respective policy scenario, in the questionnaire (order effect);

(ii) frequency of the oil spill, and respective policy scenario, in the questionnaire (*frequency effect*);

(iii) size of damage of the oil spill, and respective policy scenario, in the questionnaire (*scope effect*).

Two tests were performed:

(i) the Turnbull Ratio Test (TR test) – regarding the non parametric WTP estimates – to check possible differences in the distribution of functions;

(ii) the Wilcoxon-Mann-Whitney test (WMW test) – regarding the parametric WTP estimates – to check possible differences between average WTP across survey design specifications.

The different survey design specifications can be classified as a function of the frequency and the size of the damage (Table 11).

Table 11. *Classification of scenario's in function of the frequency and size of the damage.*

<table>
<tr><th></th><th>Light</th><th>Moderate</th><th colspan="2">Heavy</th></tr>
<tr><td>1 in 3 years</td><td>A</td><td>B</td><td colspan="2"></td></tr>
<tr><td>1 in 5 years</td><td>C</td><td></td><td colspan="2">D</td></tr>
<tr><td>1 in 10 years</td><td></td><td>E</td><td>F</td><td>G</td></tr>
</table>

A-B: Questionnaire version 1; C-D: Questionnaire version 2; E-F: Questionnaire version 3; G: Questionnaire version 4.

Order effect. The order effect can be measured by comparing two scenarios with the same frequency and the same size of damage, appearing in a different order in a questionnaire. Scenario F and scenario G can be used for this analysis. From the results of the two tests we can conclude that there is no order effect.

Frequency effect. The frequency effect can be examined for:
- (i) scenario 1/3 and scenario 1/5 (A and C)[13];
- (ii) scenario 1/3 and scenario 1/10 (B and E)[14];
- (iii) scenario 1/5 and scenario 1/10 (D and F)[14].

From the results of the TR test we can conclude that there is no frequency effect. From the results of the WM test we can conclude that there is a frequency effect between scenario 1/3 and scenario 1/10, this for the sample with protest answers. No conclusions can be drawn for the sample without protest answers.

Scope effect. The scope effect can be examined for:
- (i) light scenario and Moderate scenario (A and B);
- (ii) light scenario and Heavy scenario (C and D) [14];
- (iii) moderate scenario and Heavy scenario (E and F) [14].

From the results of the TR test we can conclude that there is no scope effect. From the results of the WM test, sample with protest votes, we can conclude that there is no scope effect. From the results of the WM test, sample without protest votes, we can not make a conclusion about a possible scope effect.

[13] Scenarios B and D cannot be used since there is also a difference in the size of damage.
[14] These scenarios can be compared since there is no order effect.

The results of the comparative analysis are not clear. Possibly, more observations per scenario may have resulted in clearer results.

4.3. Valuation function

4.3.1. Construction of the multivariate WTP model

A valuation function checks which factors influence the willingness to pay. A logit-regression model is used to build up the valuation function (see paragraph 4.2.1.). For the valuation function, the two scenarios per questionnaire are used as two separate observations with an extra variable in the function which states if the scenario appears first or second in the questionnaire. One type of questionnaire contained only one scenario (Heavy-1/5). This type of questionnaire is not used for the valuation function because for all other questionnaires the respondent has two observations. In this way every respondent gets the same weight in the valuation function. Protest answers were also excluded for calculation of the valuation function. Finally 357 (of the 571) questionnaires are used as an input for the valuation function (= 714 observations). From the econometric point of view, the building of the empirical model is characterized by the use of the 'forward step procedure'. In other words, the model starts with a constant, bid (start bid), bidlow (lower follow-up bid) and bidhigh (higher follow-up bid). A variable is added each time to the existing model. The 'Likelihood Ratio Test' examines if the new variable has an influence on the WTP. The test is done at a 5 % significance level.

4.3.2. Final model estimates and interpretation

According to our estimation results, table 13 shows the model specification that best fits the data. The coefficients in Table 13 can be interpreted as follows: a positive coefficient (or exponent of this coefficient > 1) [15] means a positive influence of the variable on the WTP. A short explanation per variable is given below.

As we can see the older the respondent the less she/he is willing to pay. In the same way, the higher the reported income of the respondent the more she/he is willing to pay.

The respondent is willing to pay more for the second programme than for the first programme. This is logical because the second programme prevents more damage than the first programme (and as a result the bid amounts are also 20 % higher in the second scenario). From the comparative analysis in paragraph 4.2.3. we can conclude that there is no order effect and for the scope effect test no conclusions could be drawn. From the results of the valuation function we can conclude that there is no scope effect, since none of the variables related to damage and measures is

[15] With exception of the variable 'Bid amount'. HANEMANN et al. (1991) states that the negative value of the coefficient need to be taken.

selected in the model. But both effects together (order effect and scope effect) seem to have an influence on the WTP[16].

The province in which the respondent resides influences the willingness to pay. The values need to be interpreted as a function of the reference province, this is the coastal province West -Vlaanderen (coefficient 0). A positive coefficient (Vlaams-Brabant, Antwerpen en Luxemburg) means that in these provinces the WTP is higher than in West-Vlaanderen.

Respondents who know the nature reserve 'de IJzermonding' (C1) are willing to pay more for the programmes. With this question we tried to select the respondents with specific nature knowledge. In addition, respondents who frequently visit the coast for the fresh and healthy air (*E32*) are prepared to pay less for the programme than the other respondents.

The estimation results regarding watching movies or reading books about nature *(E7)* need to be interpreted in function of the reference coefficient (watching and reading a lot). So, the more one watches nature films or reads books about nature the higher the willingness to pay. Equally interesting is to observe that respondents who already donated to an environmental organization or a specific environmental project are willing to pay more than other respondents. This may signal the presence of warm glow, *i.e.* the sense of moral satisfaction provided by the act of giving.

[16] For the formulation of the valuation function 714 observations are used, compared to approximately 260 observations in the difference analysis without protest answers.

Table 13. *Selected variables in the final valuation function.*

Variable	*Coefficient*	*Stand. Dev.*	*T-Stat.*	*Exp (Coeff.)*
Constant	2.2453	0.7841	2.8640	9.4432
Bid Amount	-0.0007	0.00003	-20.3300	1.0007
Order of appearance of the scenario	0.4086	0.1580	2.5860	1.5048
Living in province (reference is West-Vlaanderen = Coastal province)				
Oost-Vlaanderen	-0.7459	0.3035	-2.4580	0.4743
Vlaams-Brabant	0.0733	0.3594	0.2040	1.0761
Antwerpen	0.0163	0.3109	0.05253	1.0165
Limburg	-0.0841	0.3771	-0.2231	0.9193
Waals-Brabant	-0.8948	0.5161	-1.7340	0.4087
Henegouwen	-0.8517	0.3528	-2.4140	0.4267
Luik	-0.2152	0.3487	-0.6171	0.8064
Namen	-0.2808	0.4329	-0.6485	0.7552
Luxemburg	0.4715	0.8858	0.5322	1.6024
Brussels Region	-0.9468	0.8322	-1.1380	0.3880
Knowledge of the nature reserve 'de IJzermonding'	0.4929	0.1959	2.5160	1.6370
Visiting the coast for the fresh and healthy air	-0.4305	0.2242	-1.9200	0.6502
Watching movies or reading books about nature (reference is very often)				
Often	-0.2572	0.2206	-1.1660	0.7732
Sometimes	-0.5292	0.2339	-2.2620	0.5891
Rarely	-0.2941	0.3390	-0.8675	0.7452
Never	-1.6275	0.5155	-3.1570	0.1964
Donating to an environmental organization or a specific environmental project	1.2133	0.2649	4.5810	3.3646
Age	-0.0181	0.5916	-3.0520	0.9821
Income (reference is class 0–496 euro/month)				
496 - 992	1.3513	0.6185	2.1850	3.8623
992 - 1488	1.9996	0.6216	3.2170	7.3863
1488 - 1984	2.5384	0.6218	4.0820	12.6598
1984 - 2480	2.2105	0.6277	3.5210	9.1206
2480 - 2976	3.3619	0.6574	5.1140	28.8443
2976 - 3472	2.4854	0.6941	3.5810	12.0063
3472 - 3968	1.8171	0.7078	2.5670	6.1541
> 3968	4.2946	0.7663	5.6050	73.3037

5. SYNTHESIS

This chapter has offered an economic assessment of the loss of non-use values resulting from different types of oil spill along the Belgian Coast. For this economic assessment the contingent valuation method is used. The results show that if no policy action is undertaken to prevent oil spill damage off the Belgian coastline a significant welfare loss may result. Only taking into account the non-use values from the CV study, the welfare loss amounts up to 606 million euro, which corresponds to 0.24 % of the Belgian GDP measured at market prices for the year 2001. When we use the most conservative value (scenario with lowest damage, including protest votes and assuming that respondents that refused to cooperate have a zero WTP) then the welfare loss amounts to 120 million euro, which corresponds to 0.05 % of the Belgian GDP. The estimated welfare loss ranges less in function of the size and the frequency of the damage. Some years ago the Belgian Government has bought oil spill combating equipment for approximately 1.25 million euro (VLIZ 2001). So we can conclude that such an investment, preventing even oil spills with a relative small impact on the marine environment, can be clearly defended from a cost-benefit perspective since it costs far less than 120 million euro.

REFERENCES

- ARROW, K., SOLOW, R., PORTNEY, P.R., LEANER E.E., RADNER, R. & SCHUMAN H. (1993), *Report of the NOAA Panel on Contingent Valuation*, Federal Register 58, 4.601-4.614.

- CARSON, R.T., MITCHELL, R.C., HANEMANN, M.W., KOPP, R.J., PRESSER, S. & RUUD, P. A. (1992), *A contingent valuation study of lost passive use values resulting from the Exxon Valdez oil spill. A report to the Attorney General of the State of Alaska*, 127 p.

- CARSON, R.T., CONAWAY, M.W., HANEMANN, M.W., KROSNICK, J.A., MARTIN, K.M., MCCUBBIN, D.R., MITCHELL, R.C. & PRESSER, S. (1996), *The value of preventing oil spill injuries to natural resources along California's Central Coast*, San Diego, Natural Rescurce Damage Assessment Inc.

- CENTRUM VOOR ECONOMISCHE STUDIËN K.U.LEUVEN (2000), *Economische Waardering van Bossen, een case study voor Heverleebos-Meerdaalwoud,* VLINA96/06, 212 p.

- HANEMANN, M.W., LOOMIS, J. & KANNINEN, B. (1991), Statistical efficiency of double-bounded dichotomous choice contingent valuation, *American Journal of Agricultural Economics*, 73(4), 1.255-1.263.

- LOOMIS, J.B & WALSH, R.G. (1997), *Recreation Economic Decisions: Comparing Benefits and Costs*, Pennsylvania, USA.

- MITCHELL, R.C. & CARSON, R.T. (1989), *Using surveys to value public goods: the contingent valuation method,* Washington, D.C., Resources for the Future, 463 p.

- NUNES, P.A.L.D. (2000), *Contingent valuation of the benefits of natural areas and its warmglow component*, Leuven, KUL, 279 p.

- NUNES, P.A.L.D. (2002), *The contingent valuation of natural parks: assessing the warmglow propensity factor, new horizons in environmental economic series*, Edward Elgar Publishing, UK.

- TURNBULL, B.W. (1976), The empirical distribution function with arbitrarily grouped, censored and truncated data, *Journal of the Royal Statistical Society*, B38, 290-295.

- VAN BIERVLIET, K., BOGAERT, G., DECONINCK, M., BOGAERT, S., LE ROY, D., NUNES, P.A.L.D. AND BIOMATH (2002), *Beoordeling van mariene degradatie in de Noordzee en voorstellen voor een duurzaam beheer. Thema 1: Socio-economische kostprijs van degradatie van het mariene milieu. Taak 1.2: Ontwikkelen van socio-economische beoordelingscriteria die het maken de kostprijs van deze degradatie objectief te bepalen.*

- VAN HECKE, E. (1998), Actualisering van de stedelijke hiërarchie in België, *Tijdschrift van het Gemeentekrediet*, 52ste jaar, nr. 205, Brussel, Groep Dexia.

- VLIZ (2001), Kustverdediging: de strijd tegen de zee, *De Grote Rede*, December 2001,

- YAMANE, T. (1967), *Elementary sampling theory,* Englewood Cliffs, Prentice Hall, Inc., 405p.

ANNEX 1: QUESTIONNAIRE EXAMPLE VERSION 3 'MODERATE-SEVERE'

A. Introduction

A.1	Number of questionnaire: ____

A.6	Interview date: ___/___/___
A.7	Start interview time: ___ h ___

B. General questions

I'll start with some general questions.

B.1	Belgian citizens have different views about a number of actual social problems in our country. I'm going to name some of these social problems. Please tell me the three problems that, according to you, should be dealt with in the first place? Show card 1		
	B.1.1	Unemployment	0
	B.1.2	Environmental pollution	0
	B.1.3	Insecurity on the street	0
	B.1.4	Tax pressure	0
	B.1.5	Intolerance	0
	B.1.6	Aids	0
	B.1.7	Political chicanery	0
	B.1.8	Declining values and standards	0
	B.1.9	Use of drugs	0
	B.1.10	Uncertainty of the pensions	0

We will now go on with a question dealing specific with the theme of environmental problems.

B.2	I'll name six environmental problems, caused by humans. Please tell me the two problems that, according to you, should be dealt with in the first place? Show card 2		
	B.2.1	Air pollution caused by cars and factories	0
	B.2.2	Too much waste	0
	B.2.3	Fewer nature reserves	0
	B.2.4	Oil spills on the sea	0
	B.2.5	Pollution of rivers	0
	B.2.6	Noise caused by industries, aeroplanes	0

The government has different programs running to deal with the different social problems, including environmental problems.
In this interview we ask your opinion about **a program for the prevention of environmental damage caused by oil pollution along the Belgian Northsea Coast**. I'll start by giving you some background information.

C. Presentation of the Belgian North Sea Coast

Show card 3, North Sea

☞ The area in the North Sea for which Belgium is responsible runs approximately 60 km in sea. This survey will only deal with this part.

Show card 4, three functions

The Belgian part of the North Sea has three important functions:

☞ An **economic function**, with fishery, the extraction of sand and gravel, transportation of goods, running of hotels, restaurants and pubs and rental of apartments as the principal activities;

☞ A **recreational function** with sunbathing, walking and water sports as most practised activities;

☞ And a **unique nature function**: protected sea and coastal birds have their breeding ground and their wintering place at the Belgian Coast. Sandpipers, different fish species, lobsters, shrimps and a lot of little animals live there between a big abundance of species of plants.

Show card 5, sandbanks and beach

Off the coast, in the sea, different sandbanks lies a sort of 'sand hill landscape' under water. These sandbanks have an important nature value for a lot of species:

☞ They are the **breeding ground** for a lot of species of fishes, lobsters and shrimps.

☞ This wealth serves as the **food** for bigger fish.

☞ They are also an important source of food for seabirds. Approximately 70,000 **seabirds winter** in the area of the sandbanks. A number of these seabirds are vulnerable species.

Show card 6, photograph of birds

☞ Among other things the Common Scoter, the Red-throated Diver, the Little Gull, the Great Crested Grebe, the Guillemot and the Razorbill are protected species.

Show card 7, sandbanks and beach

☞ The parts of the beach that are flooded by the sea are a breeding ground for flatfish and sandpipers.

Show card 8, Zwin and IJzermonding

☞ The nature reserves the **Zwin** and the **IJzermonding** are both regularly flooded by the sea.

☞ That's why the Zwin and the IJzermonding contain a **great wealth of species of plants and animals** that don't occur elsewhere along the coast.

☞ For a lot of bird species these nature reserves are the pre-eminent **breeding ground**. Migratory and winter birds come there to rest and to overwinter.

C.1	Were you informed about the unique nature value of the nature reserve the IJzermonding?			
	(1)	Yes	0	
	(0)	No	0	
	(99)	No answer	0	

This natural abundance is being threat by humans: too much fishery, pollution with environmental dangerous substances and disturbance of the natural areas are only some of the dangers for the Belgian North Sea Coast. Also accidents with ships and tankers can cause oil pollution and can threaten this natural abundance.

C.2	Have you heard or read something in the news about accidents with oil tankers, in recent years?			
	(1)	Yes	0	[C.3]
	(0)	No	0	
	(99)	No answer	0	

C.3	What exactly did you hear or read?		
	(1)	Erika before the French Coast	0
	(2)	Jessica, Galapagos Islands (Ecuador)	0
	(3)	Others, to specify	0
	______________________		
	______________________		
	(99)	No answer	0

Along the Belgian Coast a number of accidents have already happened, but the damage to nature was small in comparison to the damage in the surrounding countries. Accidents with large effects happen regularly in nearby coastal zones: France and Wales have been hit several times.

Along the Belgian North Sea Coast with its narrow waterway, frequent fog and bad weather conditions, there is almost certainly going to be an accident involving an oil tanker in the next ten years, where also the nature of the Belgian North Sea Coast will get damaged.

To prevent oil damage to nature:

- All oil tankers in European waters should have two outer hulls instead of the single hull most of them have now.
- The control of the ships in the ports, also oil tankers, should be much more stringent.

Show card 9, Time axis

☞ International law will require these measures from the year 2010. From the year 2010, with these international measures the chance of an oil pollution incident with severe consequences for the surrounding environment will be very small.

☞ To avoid such an accident **meanwhile, a temporary prevention and intervention program** can be put into action by the Belgian government.

In this way, the nature value of the Belgian part of the North Sea is protected in the best possible way.

Show card 9, Time axis
Show card 10, Financing

To be able to finance the Belgian programme, a financial contribution is expected from the producers and the consumers of oil products.

From the **producers of oil**, the oil companies, major efforts are expected:

☞ They have to pay higher compensation when accidents happen or when they violate the law and they have to contribute more in a disaster fund. In that way more money is available in a disaster fund. With this money all operational costs of the intervention programme can be paid, when an accident happens.

A financial effort is also expected from the **consumers of oil**, the Belgian citizens. Crude oil is incorporated in a lot of products that everybody uses daily: fuel for heating and transportation, medicines, plastics, shampoo, nylon, etc. So almost every Belgian citizen is a consumer of oil.

☞ Money is also needed to finance the investments of the intervention- en prevention programme. That's way a North Sea fund is established, managed by the government. The effort of every Belgian family would consist of a one time financial contribution into this North Sea fund. Every family should pay this contribution in September of this year.

It's also important that the money of the North Sea fund can only be used for the execution of the programme. The government cannot use the money for other investments.

The oil companies would pay for the operational costs of the intervention programme and the Belgian citizens would pay for the investments of the prevention and intervention programme.

Show card 11, Referendum

But! It's not sure that this programme will be implemented.

- ☞ The programme will only get through if more than half of the Belgian population is prepared to pay for it. And if that is the case, everybody will have to pay the contribution in September of this year.
- ☞ But if more than half of the Belgian population refuses to pay for the programme, the programme will not be implemented.

Through this survey we want to know if the Belgian population would say 'yes' or 'no' to the prevention and intervention programme, if a referendum were held. The content of the programme and the size of the contribution will be dealt with later in the survey.

C.4	Have you any questions about what I've just said?			
	(1)	Yes	0	[C.5]
	(0)	No	0	
	(99)	No answer	0	

C.5	Which?
	____
	____
	____________ (Repeat the part where the problem is situated or see below.)

C.6 If there is a remark about the contribution of the oil companies, tick here □ and explain again:

The government cannot legally oblige the oil companies to pay for all the costs of the programme. What the government can do, is to oblige the oil companies to pay more compensation when accidents happen or when they violate the law. The government can also oblige the producers of oil to contribute more in a disaster fund. The cleaning costs of an accident can be paid with this money. These payments cannot be charged to the consumer.

These two extra measures make sure that one of the polluters, the oil companies, put in more effort than before. The other polluters, the ship owners are unfortunately numerous and unknown (foreign ship owners, captains and the crew of ships in transit). All these ships discharge oil into the North sea, for example when they clean their ships. It's very difficult to take legal action against them.

D. Scenario's

I've already told you that there is almost certainly going to be an accident involving an oil tanker in the next ten years, where also the nature of the Belgian North Sea Coast will be damaged by oil. We can only not predict how severe the damage will be. I'll present you two accidents that both cause certain damage to nature. The damage in the second accident is more serious than in the first one. We'll ask to evaluate both scenarios separately.

Scenario A

Show card 12, Scenario A

- ☞ Suppose there's an accident in which 5,000 m^3 oil lands into the sea. What are the consequences for the Belgian coastal zone.
- ☞ Approximately 20,000 seabirds would die because of oil pollution, which is 30 percent of the present population.
- ☞ 10 percent of the present fishes, crabs, shrimps and lobsters would die by poisoning.
- ☞ 25 km of the sandy beach would get polluted with oil, which is approximately 40 percent of the total length of the Belgian Coast.
- ☞ The nature reserve the IJzermonding would become polluted with oil.
- ☞ The Zwin would never be threatened. Within 10 years all damage will restore naturally. No animal species will become extinct and no area will get permanent damaged.

Take another look at card 12.

Show card 12, Scenario A
Show card 13, Programme A

The damage to nature can be avoided with a prevention and intervention programme:
The prevention part, which should prevent happening such accidents, consist of two specific measures:

- ☞ As a first measure, 10 km separate 'straits' would be marked for oil tankers. This means that oil tankers can only sail in a special zone marked by light buoys. In that way the usual ship traffic is completely separated from oil tanker traffic.
- ☞ The second measure consists of an extension of the radar control with a permanent and continuous communication system. This system is comparable to the communication system of an airport.

The programme doesn't only consist of measures needed to prevent such accidents. The programme also has an intervention part, which needs to minimise the damage when such an accident should nevertheless happen. This part of the programme consists of three specific measures:

- ☞ First of all two big tugboats would be purchased. When an accident happens, or when a ship gets in trouble the tugboats can immediately be called in. The tugboat can bring the ship that loses oil or that threatens to lose oil to safer places. In that way a tugboat can ships losing more oil after an accident than when they are floating in the sea
- ☞ The second measure consists of the purchase of two oil-combating platforms for oil fighting on the sea. When oil gets into the sea, the two 'arms' of the platform can shovel up the oil out of the water.
- ☞ The third intervention measure consists of a specific measure to close the nature reserve the IJzermonding off from the sea.

D.1.A	Have you any questions about what I've just said?			
	(1)	Yes	0	[D.2.A]
	(0)	No	0	
	(99)	No answer	0	

D.2.A	Which?
	______________________________
	______________________________
	______________ (Repeat the part where the problem is situated)

You should pay a one-time contribution to the North Sea fund in September of this year, if this programme is approved by more than half of the Belgian population. Think about it well before you answer these questions.

D.3.A Would you vote in favour or against the programme if it would cost your family a one time contribution of 24.79 Euro. Please think about this well, taking into account the current household income and expenditures.

				Revision	
(1)	In favour	0	[D.4.A]	**0**	**[D.4.A]**
(0)	Against	0	[D.5.A]	**0**	**[D.5.A]**
(99)	No answer	0	[D.5.A]	**0**	**[D.5.A]**

D.4.A Would you vote in favour or against the programme if it would cost your family a one time contribution of 61.97 Euro instead of 24.79 Euro. Please think about this well, taking into account the current household income and expenditures.

				Revision	
(1)	In favour	0	**[Scenario B]**	**0**	**[D.10]**
(0)	Against	0	**[Scenario B]**	**0**	**[D.10]**
(99)	No answer	0	**[Scenario B]**	**0**	**[D.10]**

D.5.A Would you vote in favour or against the programme if it would cost your family a one time contribution of 9.92 Euro instead of 24.79 Euro. Please think about this well, taking into account the current household income and expenditures

				Revision	
(1)	In favour	0	**[Scenario B]**	**0**	**[D.10]**
(0)	Against	0	[D.6.A]	**0**	**[D.6.A]**
(99)	No answer	0	[D.6.A]	**0**	**[D.6.A]**

D.6.A Can you give the most important reason why your household is not prepared to pay this one time contribution? Show card 14

			Revision	
(1)	I don't believe in the proposed project	0	**0**	**[D.10]**
(2)	The proposed project isn't worth that much	0	**0**	**[D.10]**
(3)	The oil companies have to pay all the costs	0	**0**	**[D.10]**
(4)	I don't want to pay something extra, the government has to pay everything	0	**0**	**[D.10]**
(5)	My income doesn't allow me to pay this amount	0	**0**	**[D.10]**
(6)	Others, specify	0	**0**	**[D.10]**

____
____
____

Scenario B

Now a second situation is presented to you. This time it's a bigger accident. More oil lands in the sea. In view of the fact that there is more oil pollution, more nature will get damaged.

Show card 12, Scenario A
Show card 15, Scenario B

Suppose that's an accident in which 10,000 m^3 oil lands into the sea. Which are the consequences for the Belgian coastal zone?

- ☞ Approximately 43,000 seabirds would die because of oil pollution, which is 65 percent of the present population.
- ☞ 20 percent of the present fishes, crabs, shrimps and lobsters would die of poisoning.
- ☞ 60 km of the sandy beach would get polluted with oil, which is approximately 90 percent of the total length of the Belgian Coast.
- ☞ The nature reserves the IJzermonding and the Zwin would become polluted with oil.
- ☞ Within 10 years all damage will restore naturally. No animal species will become extinct and no area will be permanent damaged.

Take another look at card 15 and compare the differences with the previous accident that was presented to you on card 12.

Show card 12, Scenario A	Show card 13, Programme A
Show card 15, Scenario B	Show card 16, Programme B

This damage to nature can again be avoided with a prevention and intervention programme. Because the damage is more severe than in the previous accident, more measures will be necessary.

The damage to nature can be avoided with a prevention and intervention programme:

The prevention part, consist out of two specific measures:

- ☞ As a first measure, 20 km separate 'straits' would be marked for oil tankers. This is double the first programme.
- ☞ The second measure consists of an extension of the radar control with a permanent and continuous communication system. This is the same as in programme A.

The intervention programme consists of three specific measures:

- ☞ First of all three big tugboats would be purchased. This one more than in programme A.
- ☞ The second measure consists of the purchase of three oil-combating platforms for oil fighting on the sea, instead of two in the first programme.
- ☞ The third intervention measure consists of a specific measure to close the nature reserves the IJzermonding and the Zwin off from the sea. In programme A, only the IJzermonding was closed off.

D.1.B	Have you any questions about what I've just said?			
	(1)	Yes	0	[D.2.B]
	(0)	No	0	
	(99)	No answer	0	

D.2.B	Which?
	______________________________
	______________________________
	______________ (Repeat the part where the problem is situated)

You should pay a one-time contribution to the North Sea fund in September of this year, if this programme is approved by more than half of the Belgian population. Think about it well before you answer these questions.

It's important that you try not to take account of the answers you gave in the previous scenario.

D.3.B Would you vote in favour or against the programme if it would cost your family a one time contribution of 29.75 Euro. Please think about this well, taking into account the current household income and expenditures.

				Revision	
(1)	In favour	0	[D.4.B]	**0**	**[D.4.B]**
(0)	Against	0	[D.5.B]	**0**	**[D.5.B]**
(99)	No answer	0	[D.5.B]	**0**	**[D.5.B]**

D.4.B Would you vote in favour or against the programme if it would cost your family a one time contribution of 74.37 Euro instead of 29.75 Euro. Please think about this well, taking into account the current household income and expenditures.

				Revision	
(1)	In favour	0	[D.7]	**0**	**[E]**
(0)	Against	0	[D.7]	**0**	**[E]**
(99)	No answer	0	[D.7]	**0**	**[E]**

D.5.B Would you vote in favour or against the programme if it would cost your family a one time contribution of 12.49 Euro instead of 29.75 Euro. Please think about this well, taking into account the current household income and expenditures

				Revision	
(1)	In favour	0	[D.7]	**0**	**[E]**
(0)	Against	0	[D.6.B]	**0**	**[D.6.B]**
(99)	No answer	0	[D.6.B]	**0**	**[D.6.B]**

D.6.B Can you give the most important reason why your household is not prepared to pay this one time contribution? Show card 17

			Revision	
(1)	I don't believe in the proposed project	0	**0**	**[E]**
(2)	The proposed project isn't worth that much	0	**0**	**[E]**
(3)	The oil companies have to pay all the costs	0	**0**	**[E]**
(4)	I don't want to pay something extra, the government has to pay everything	0	**0**	**[E]**
(5)	My income doesn't allow me to pay this amount	0	**0**	**[E]**
(6)	Others, specify	0	**0**	**[E]**

____

____

____

D.7	We now offer you the chance to revise your questions. This is for instance necessary if you may have understood something badly. Was it clear from what I proposed to you, that your family should pay a one-time contribution to the North Sea fund?		
	(1)	Yes	0
	(0)	No	0
	(99)	No answer	0

D.8	Was it clear from what I proposed to you, that your family should only pay for one of the two programmes?		
	(1)	Yes	0
	(0)	No	0
	(99)	No answer	0

Now, you know that you should only pay a one-time contribution for one programme, programme A or programme B.

D.9	Now that you know the content of programme B, you can maybe give a better judgement about programme A. Do you wish to revise your judgement about the first programme, programme A?			
	(1)	Yes	0	*[Revision D.3.A -> D.6.A]*
	(0)	No	0	**[E]**
	(99)	No answer	0	**[E]**

D.10	Do you wish to revise your judgement about the second programme, programme B?			
	(1)	Yes	0	*[Revision D.3.B -> D.6.B]*
	(0)	No	0	**[E]**
	(99)	No answer	0	**[E]**

E. Respondent household questions

I would like to ask some questions about your family.

E.1	Do you or anyone in your household, sometimes visit the Belgian North Sea Coast?			
	(1)	Yes	0	[E.2]
	(0)	No	0	[E.4]
	(99)	No answer	0	[E.4]

E.2	How many times have you, or anyone in your household, visited the Belgian North Sea Coast the last year? Show card 18		
	(1)	One or two times	0
	(2)	Three to ten times	0
	(3)	More than ten times	0
	(99)	No answer	0

E.3	What is the most important reason for your family to visit the Belgian Coast?
	____________________
	____________________
	____________________

E.4 If the family has a residence at the Belgian Coast, indicate here □.

E.5	Do you or anyone in your household, sometimes go on a beach, sailing or surfing vacation abroad?		
	(1)	Yes	0
	(0)	No	0
	(99)	No answer	0

E.6	Would oil pollution before the Belgian North Sea Coast have any consequences for your job or your income?		
	(1)	Yes	0
	(0)	No	0
	(99)	No answer	0

E.7	How many times do you watch television programs or do you read books about nature? Show card 19		
	(1)	Many times	0
	(2)	Often	0
	(3)	Sometimes	0
	(4)	Rarely	0
	(5)	Never	0
	(99)	No answer	0

E.8	Are you, or anyone in your family, a member of a nature/environmental organisation?			
	(1)	Yes	0	[E.9]
	(0)	No	0	[E.11]
	(99)	No answer	0	[E.11]

E.9	Of which nature/environmental organisation(s)?
	____________________
	____________________
	____________________ [E.10]

E.10	During the last year, have you paid a contribution to an environmental organisation or for a specific environmental project, on top of the amount paid for the nature/environmental organisation of which you are a member?			
	(1)	Yes	0	[E.12]
	(0)	No	0	**[F]**
	(99)	No answer	0	**[F]**

E.11	During the last year, have you paid a contribution to an environmental organisation or for a specific environmental project?			
	(1)	Yes	0	[E.12]
	(0)	No	0	**[F]**
	(99)	No answer	0	**[F]**

E.12	What amount have you paid during the last year?		
		__________ Euro	
	(99)	No answer	0

F. Control and evaluation questions

Please think very well about the problem of the risk of oil pollution off the Belgian North sea Coast that I've described you until now.

F.1	Did it seem to you that the North Sea Prevention Program I told you about would be completely effective in preventing harm from Belgian North Sea Coast oil spills, mostly effective, somewhat effective, not too effective, or not effective at all? Show card 20			
	(1)	Completely effective	0	
	(2)	Effective	0	
	(3)	Not effective at all	0	
	(99)	No answer	0	

F.2	Thinking about everything I have told you during this interview, overall did it try to push you to vote one way or another, or did I let you make up your own mind about which way you vote?			
	(1)	Pushed one way or another	0	[F.3]
	(2)	Let me make up own mind	0	**[G]**
	(99)	No answer	0	**[G]**

F.3	Which way did you think it pushed you?		
	(1)	Vote in favour of the programme	0
	(2)	Vote against the programme	0
	(99)	No answer	0

G. Socio-demographic information

To end, I would like to ask you some personal questions. Your answers will be treated confidentially. Your name or the name of your family will never be used in the study.

G.1	What is your date of birth? __/__/__		
	(99)	No answer	0

G.2	What is your professional situation at this moment? Show card 21		
	(1)	Incapable to work	0
	(2)	Retired	0
	(3)	Student	0
	(4)	Housewife	0
	(5)	Job-seeker	0
	(6)	Worker in the private sector	0
	(7)	Employee in the private sector	0
	(8)	Civil servant	0
	(9)	Self-employed without personnel	0
	(10)	Employer	0
	(11)	Assistant relative	0
	(12)	Others, to specify	0
	________________		
	________________		
	(99)	No answer	0

G.3 What is the highest degree you received? Show card 22

(1)	Primary education or no degree	0
(2)	Lower secondary education, professional or technical	0
(3)	Lower secondary education, general	0
(4)	Higher secondary education, professional or technical	0
(5)	Higher secondary education, general	0
(6)	Higher, non-university education	0
(7)	Higher, university education	0
(8)	Others, to specify	0
	______	
	______	
(99)	No answer	0

G.4 I'll present you different income categories. Can you tell me in which category the common monthly take-home pay of your family can be placed? You need to take into account the income of every family member and also unemployment benefits, pension, family allowance, social benefits and possibly other earnings.
Your answer is fully confidential. The data will only be used for statistical analysis and your name or the name of your family will never be used in the study. Show card 23
You can just indicate the number of the corresponding category.

(1) 0 – 495.79	0	(4) 1487.36 – 1983.15	0	(7) 2974.72 – 3470.51	0
(2) 495.79 – 991.57	0	(5) 1983.15 – 2478.94	0	(8) 3470.51 – 3966.39	0
(3) 991.57 – 1487.36	0	(6) 2478.94 – 2974.72	0	(9) > 3966.39	0
		(99) No answer	0		

G.5 Do you think that the family income will rise, stay equal or decline in the future?

(1)	Rise	0
(2)	Stay equal	0
(3)	Decline	0
(99)	No answer	0

G.6 How many of your children are living at home? ____

(99) No answer 0

G.7	Do you have a partner you're living with?			
	(1)	Yes	0	[G.8]
	(0)	No	0	**[H]**
	(99)	No answer	0	**[H]**

G.8	What is the date of birth of your partner?		__/__/__
	(99)	No answer	0

H. Personal characteristics

To allow my supervisor to check my work, I would like to have your first name. Your name or the name of your family will never be connected with the answers. Everything will be handled discretely and anonymously.

H.1	First name: ______________________
H.2	End of interview: __ h __

Thank you for your cooperation.

To fill in after the interview

H.3	Sex?		
	(1)	Man	0
	(0)	Woman	0

I. Control questions

I.1	Did the interview go well?			
	(1)	Yes	0	[I.3]
	(0)	No	0	[I.2]

I.2	Why didn't the interview go well?		
	(1)	Because the respondent didn't understand the questionnaire well	0
	(2)	Because the respondent wanted to know too much detail	0
	(3)	Because the respondent clearly didn't have any interest	0
	(4)	Because the survey was frequently interrupted	0
	(5)	Others, specify	0
	______________________________		
	______________________________		

I.3	Did the respondent well understand every part of the questionnaire?			
	(1)	Yes	0	[I.6]
	(0)	No	0	[I.4]

I.4	Which parts were not (well) understood by the respondent?		
	(1)	General questions	0
	(2)	Presentation of the Belgian North sea Coast	0
	(3)	Scenario A	0
	(4)	Programme A	0
	(5)	Scenario B	0
	(6)	Programme B	0
	(7)	Relation between scenario A and B	0
	(8)	WTP-questions	0
	(9)	Control and evaluation questions	0
	(10)	Socio-demographic information	0
	(11)	Person characteristics	0

I.5	Why do you think the respondent didn't understand these parts (well)?		
	(1)	Because of the questions asked	0
	(2)	From her/his comment	0
	(3)	Because she/he didn't have any interest	0
	(4)	Others, specify	0
	______________________________		
	______________________________		
	______________________________		

I.6	Do you think the respondent answered honestly and kept in mind his/her budget restriction when expressing his/her WTP?			
	(1)	Yes	0	[stop]
	(0)	No	0	[I.7]

I.7	What do you think about the expressed WTP of the respondent?		
	(1)	I suppose that the WTP is lower in reality	0
	(2)	I suppose that the WTP is higher in reality	0

Check if everything is filled in !!!!!!!!!!!!!!

PART IV.

TRADE-OFF BETWEEN ENVIRONMENTAL AND SOCIO-ECONOMIC FACTORS IN THE BELGIAN PART OF THE NORTH SEA

CHAPTER 9

THE USE OF MATHEMATICAL MODELS FOR ESTIMATING OIL POLLUTION DAMAGE AT SEA

Description of the system developed in the frame of the MARE–DASM project

Ir. SERGE SCORY

Royal Belgian Institute of Natural Sciences,
Management Unit of the North Sea Mathematical Models, Belgium

Gulledelle, 100, 1200 Brussels, Belgium
Tel.: + 32 2 773 21 33, Fax: + 32 770 69 72
E-mai: S.Scory@mumm.ac.be

1. SCOPE

Petroleum and its more or less refined derivative products are transported over all the seas of the world. Furthermore, fuel is the most widespread propulsion means and is therefore present onboard the vast majority of vessels. Although navigation, due to economic and regulatory constraints, globally evolves towards more safety, oil spills are likely to remain inevitable. Accidents will occur, due to extreme natural conditions, technical failures or human errors. No human system will ever be fully safe.

When spilled at sea oil products affect the environment. The intensity of the damage depends on several factors: *e.g.* the quantity spilled, the toxicity of the hydrocarbons, the meteorological and hydrodynamic conditions, the distance to the possible

F. Maes (ed.), Marine Resource Damage Assessment, Liability and Compensation for Environmental Damage, 211–252.

sensitive targets (sea bottom, coastline) and the time the oil spill would need to travel there, the sensitivity of the habitats and living communities that will be affected and so on.

To cope with all these parameters integrated modelling seems the better approach for assessing all possible damages. In the frame of the *MARE-DASM* project substantial attention has been paid to the development of a system of integrated models aiming at "*assessing the consequences of accidental spillage of oil at sea at the environmental and socio-economic levels*".

This task implied the improvement of existing models, the development of missing ones and incorporating the findings of two other studies performed within the project, namely the definition of socio–economic criteria for assessing the cost associated with the degradation of the marine environment due to the pollution and the refinement of a *Biological Effect SubModel.*

The present chapter describes the processes involved when oil is spilled on the sea and the various submodels that were coupled. An example of a possible application of the integrated system is also given.

2. PROCESSES AND MODELS

Assessing the consequences of an accidental release of a pollutant at sea implies considering a chain of information and processes:

(i) the event occurs in given environmental conditions (meteorology, sea state, ...);
(ii) the pollution has its own specific characteristics (type of pollutant, quantity, location of the release, ...);
(iii) it occurs in a specific natural and human environment;
(iv) the intrinsic characteristics of the pollution evolve over space and time, and so do the induced effects;
(v) resources (*i.e.* living or non–living resources and human activities that use them) that occur to be hit by the pollution might be affected instantaneously or on the long term;
(vi) cross-connections between resources might induce indirect effects;
(vii) ...

To meet the objective of an integrated system, several "models" have thus to be considered. Their integration further implies that these models run in a coherent functional framework, with consistent time and space scales, using the appropriate variables to feed downstream modules.

Conceptually, the set of models is divided as shown in Figure **1**. The various models under consideration are described hereafter, together with a description of the relevant processes.

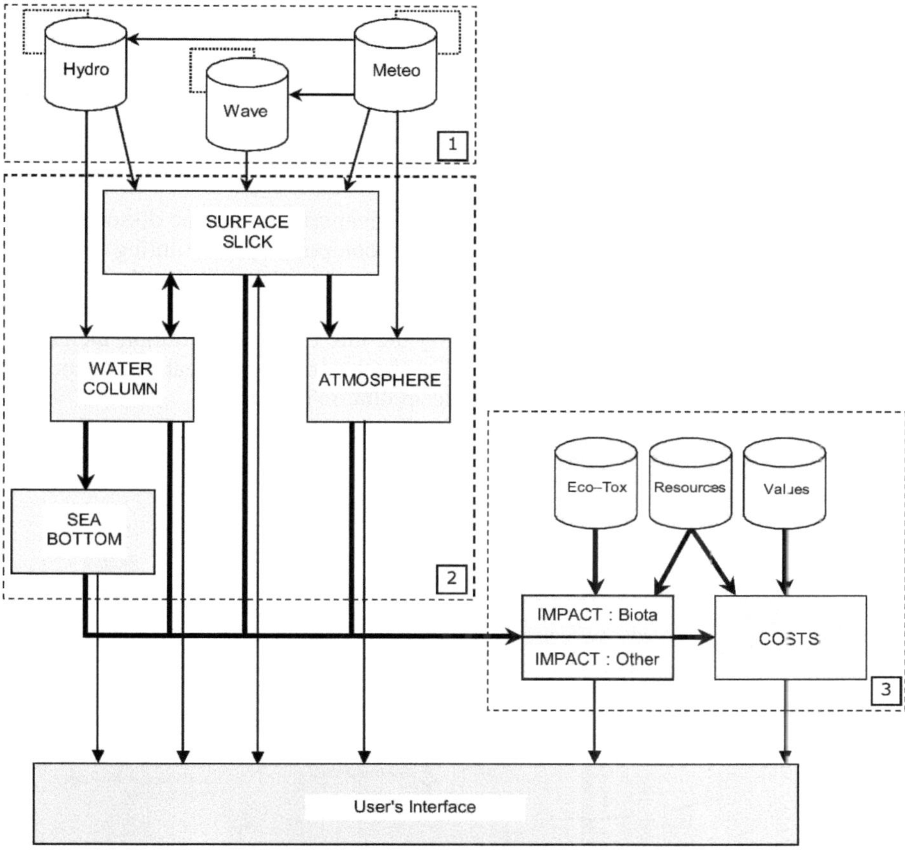

Figure 1. *Reference conceptual scheme of the integrated MARE–DASM system of models. 1: Environmental conditions; 2: Physical–chemical behaviour of the pollutant; 3: Effects, impacts and quantification.*

2.1. General behaviour of oil spilled at sea

The balance of mass and the physical characteristics of oil spilled at sea (and of most of the other liquid pollutants) evolve under the combined influence of endogenous processes and environmental constraints.

The main processes are shown in Figure 2. An oil slick drifts under the combined influence of the wind and the surface currents. It spreads due to endogenous processes (spreading forces like gravity and surface tension, damped by inertia, the viscosity of the fluid and the stress at the oil/water interface) and turbulence processes (oil droplets that have entered the water column due to wave movements drift at slower velocity that the surface slick and they resurface – due to their buoyancy – farther with respect to the centre of the slick than if they had remained at the surface). The volume of oil remaining at the sea surface decreases with time due to the evaporation of its lighter compounds, the dissolution of its soluble compounds, the adsorption of oil to particles or when part of the pollution hits the coast.

Oil at sea “weathers” rapidly, mainly due to the evaporation and the dissolution that modify the relative proportion of the various compounds constituting the “oil”. These processes affect the apparent physical properties of the oil, *i.e.* the properties of the oil considered as a homogeneous product. The properties that are most of the time taken into consideration when studying the fate of an oil pollution incident at sea are the apparent density and viscosity, as these parameters greatly influence the effectiveness of some of the possible counter-pollution measures.

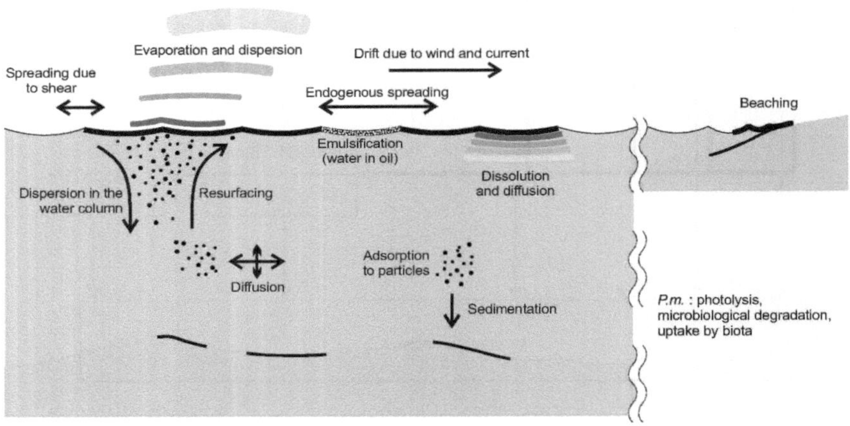

Figure 2. *Major natural processes that affect the repartition of oil in the marine environment.*

Some oil products also show a tendency to form an emulsion with water (water in oil), resulting in higher values of the apparent gravity and viscosity. Compared to the oil alone, such an emulsion can be considered as stable: light and soluble compounds are “trapped” inside and, consequently, evaporation and dissolution slow down.

Some other processes affecting the fate of an oil slick should be mentioned for completeness: photolysis, microbiological degradation, uptake by biota, … As they are most of the time of several orders of magnitude weaker (at least in the early life of a spill – up to a fortnight) they were not considered further in this study.

Finally, human actions (counter-pollution measures at sea) might influence the "natural" processes described above. These actions either increase or decrease the natural processes (mechanical increase of turbulence, chemical dispersion, ...) and accordingly influence the characteristics of the pollution, such as its extent and the repartition of mass between the various compartments. They can also directly influence the extent of the polluted area (use of booms, ...) and the global mass balance (skimming, pumping, ...).

2.2. Environmental conditions

The basic environmental conditions that are relevant for modelling oil spill behaviour and impact in temperate seas are: wind velocity and direction, (surface) current velocity and direction, air and sea temperatures. These parameters are most often provided by the operational forecasting systems. In the present application, the meteorological conditions (wind intensity and direction) are forecasts issued by the UK Met. Office. The currents are provided by a 2D storm-surge model. For the purpose of this study, the extrapolation from the depth-averaged current to the surface current is handled by the "wind transfer coefficient" in the advection module of the oil spill model. Sea state (which is used in the oil spill model as a measure of the surface turbulence intensity), at last, is given by a second-generation wave model[1].

For these environmental conditions, the word "forecast" should not be misunderstood. Although MUMM's "operational forecasting system" is primarily designed to run in operational mode – *i.e.* twice a day, following the common updating scheme of the meteorological forecasts – it is also able to provide information for a given elapsed period of time ("hindcast"). In this case, it normally uses meteorological information reconstructed from analysed fields (gridded observations).

Furthermore, the oil spill model can handle meteorological forecasts provided by the user as a time series of wind intensity and direction. In this case, however, the currents – if used for describing the fate of the pollution – are not guaranteed to be consistent with meteorological conditions.

2.3. Oil spill modelling

Various techniques have been developed to model the physical and chemical processes involved in the fate of a marine oil spill.

Table **1**, adapted from REED *et al.*[2], gives an overview of these techniques and a short description of the corresponding techniques in our reference model (MU–SLICK).

---

[1] VAN DEN EYNDE, D., SCORY, S. & MALISSE, J.-P. (1995), Operational modelling of tides and waves in the North Sea on the Convex C230 at MUMM, in: *Proc. European Convex Users' Conference 1995*, Brussels, Belgium, 24-27 October 1995.

[2] REED, M., JOHANSEN, Ø., BRANDVIK, P.J., DALING, P., LEWIS, A., FIOCCO, R., MACKAY, D.

***Table 1.** A synthetic comparison between the present state of the art in oil spill modelling (REED et al.) and the MU-SLICK model. "–", "~" and "+" indicate in a schematic way how the authors rank the methods with respect to their ability to yield useful results.*

	Processes		***Methods***	
	"State of the Art"	*MU-SLICK*	*"State of the Art"*	*MU-SLICK*
Advection	3D: Influenced by surface current, wind & waves, vertical turbulent mixing & buoyancy.	2D: Influenced by surface current & wind. Vertical transport not explicitly taken into account.	▪ Direct or indirect linking to hydrodynamical models (+); ▪ Current atlases or static approximation (–).	▪ Indirect linking to a 2D hydrodynamical model, provision for manual data entry.
Spreading	Due to gravity & surface tension, limited by inertia and viscosity	Due to gravity & surface tension, limited by inertia and viscosity	▪ Classical Fay & Hoult equations (~), ▪ Inclusion of effects of shear spreading and other environmental stresses (+). ▪ Spreading should be linked to vertical dispersion.	▪ Differential equation based on FAY's concept. ▪ MU–SLICKLETS has the potential to take environmental stresses and vertical dispersion into account.
Evaporation	Rate depends on: oil characteristics, oil thickness & environmental conditions (t°, wind speed, ...)	Rate depends on: oil characteristics, oil thickness & wind speed.	▪ Pseudo component method (–), ▪ Analytical method (–), ▪ Empirical correlation (–).	▪ Empirical correlation
Natural dispersion	Rate depends on: oil characteristics, oil thickness and sea–state.	Rate depends on: oil thickness and sea–state.	Dispersion rate as a function of oil type, wave energy and %–age of breaking waves, according to DELVIGNE & SWEENEY. (~+)	Dispersion rate proportional to significant wave height and volume of oil available for dispersion.

Table **1** *(cont.)*

& PRENTKI, R. (1999), Oil Spill Modeling towards the Close of the 20th Century: Overview of the State of the Art, *Spill Science & Technology Bulletin*, 5 (2), 3–16.

Table 1. *(cont.)*

	Processes		*Methods*	
	"State of the Art"	*MU-SLICK*	*"State of the Art"*	*MU-SLICK*
Emulsification	Water–in–oil emulsion.	Water–in–oil emulsion.	▪ Rate as a function of maximum water content and water uptake rate scaled to sea state. (~) ▪ Prediction based on oil composition should be available in the near future. (+)	▪ Rate as a function of maximum water content and significant wave height.
Oil–ice interaction	Complex phenomenon, due to the many oil–ice interactions and the difficulty to compute ice formation at scales valid for oil spill modelling.	N/A		N/A But the MU–SLICK model, with a few modifications, has been coupled to a sea–ice model of Weddel Sea.
Oil–shoreline interaction	Interaction as a function of oil viscosity, tide level and type of sediments	A spill that hits the coast does not move anymore	Mass remaining ashore calculated as a first–order process, with a constant or variable removal rate. (–)	When the centre of mass hits a boundary of the "wet" domain it stops. MU–SLICKLETS allows for a more detailed behaviour description.
Oil properties	Changes in density, viscosity, pour point, flash point, ..., depending on t°, evaporation, water content,...	Changes in density, viscosity & surface tension, depending on weathering processes.	Simple extrapolation formulae (~), Empirical data (+).	Empirical relationships, coupled with the set of equations describing the weathering processes.
Spill response	To reduce the environmental impact of spills to improve the selection of response strategies.	Allows to take mechanical recovery and chemical dispersants action into account;	Not often implemented	A net removal rate can be imposed, corresponding to the mechanical recovery rate, The natural dispersion rate can be increased, to account for the use of chemical dispersants.

The surface behaviour model used in this study is an improved version of the "MU–SLICK" model[3], called "MU–SLICKLETS". The initial model is a deterministic model involving transport, spreading and weathering processes. Adapted from KUIPERS[4], it uses the standard approach of describing separately the drift of the slick and the evolution of its shape and its physical and chemical characteristics. It is based on the concept of a circular (cylindrical) slick, spreading under the influence of the endogenous forces (gravity, viscosity, inertia and net surface tension at its edge) and drifting as a whole under the combined influence of the wind and the currents. As this model has always been coupled to a 2D storm-surge model, providing for depth-averaged currents, the difference between the mean current and the surface current that actually drives the oil is modelled through an increased wind drag coefficient in the formula giving the drift velocity of the oil.

This model includes functions to describe the major processes described in the previous section:

(i) evaporation, as a function of the surface of the slick, of the wind speed, of the percentage of emulsified oil and of a *a priori* splitting of the initial oil volume into an evaporable and a non–evaporable fraction,

(ii) vertical dispersion, as a function of the volume of the slick, of the percentage of emulsified oil and of the sea surface turbulence (characterized by the significant wave height),

(iii) emulsification, as a function of the surface of the slick, of the percentage of non–emulsified oil and of the sea surface turbulence (characterized by the significant wave height),

(iv) dissolution in water, as a function the surface of the slick.

The time evolution of the apparent gravity, viscosity and net surface tension are also computed, consistently with the changes induced by the processes mentioned here above.

Although it has many times proven to give valuable and sufficient information to cover the needs of people in charge of setting up counter-pollution measures in case of actual accidental spills, this model suffers several limitations with respect to the needs of the MARE–DASM project.

The major limitations that were identified were mainly related to the unrealistic shape of the oil slick, leading to a lack of precision about what was actually

---

[3] SCORY, S. (1982), Etude du déplacement et de l'étalement d'une nappe d'hydrocarbures déversée en mer, sous l'effet du vent, des courants et de la houle. Travail de fin d'études, Université de Liège (in French); SCORY, S. (1991), The MU–Slick Model, MUMM Report, CAMME/91/03, 28 p. ; SCORY, S. (1995), Models used by the Belgian authorities in case of accidental spills at sea, Paper presented at the workshop on "Prediction of Short-Term Transport and Dispersion of Acidental Spills from Shipping and Off-Shore Industry, The Hague, The Netherlands, 15-17 November 1995, 5 p.

[4] KUIPERS, H. D. (1981), *SMOSS - A simulation model for oil slicks at sea,* Delft University of Technology and North Sea Directorate, Ministry of Transport and Public Works, Rijswijk, The Netherlands.

happening and where. The interaction with the coastline, especially, could not be described satisfactorily.

It has therefore be decided to move to a hybrid model, taking advantage of the functions that proved to give satisfactory results but modelling the slick as an ensemble of particles having their own characteristics and behaviour. The "MU–SLICKLETS" model iterates between the two formalisms: most of the processes are modelled as advective processes (horizontal drift under the influence of the current and of the wind, endogenous spreading horizontal spreading due to turbulence, dispersion in the water column and resurfacing) that affect each of the particles individually. Concurrently, at each time step, a circular slick corresponding to the characteristics (relative positions and individual volumes) of the particles present at the surface is defined and reference rates for evaporation, dissolution, dispersion, emulsification are computed. These rates, in turn, are used to define the statistical distribution to randomly select the particles that, during the next step, will either evaporate, dissolve, disperse or mix with water in a stable emulsion.

A particle that has been "selected" to enter the water column due to the vertical mixing induced by the sea surface turbulence goes down up to a depth defined as a function of the significant wave height. If not selected as a particle that eventually sinks to the bottom (due to its adsorption to suspended materials, for instance), it drifts vertically, according to its buoyancy, and horizontally, at a speed that takes into account the diminishing influence of the wind stress on the water current with depth.

The corresponding behaviour generates patterns that "look" closer to actual ones(see Figure **3**). PIt is however difficult to compare the results of the model with observations of actual spills in a more quantitative manner. Nevertheless, it provides much more useful and precise information when dealing with interactions with "targets" that are represented by linear elements (coastline), ensuring a more adequate assessment of the impact thereon.

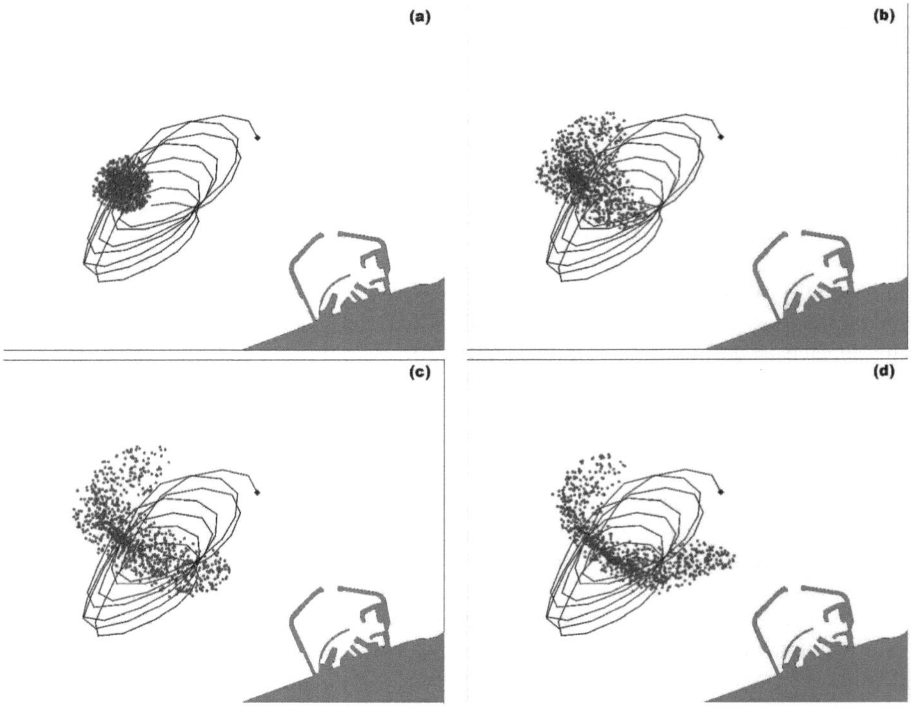

Figure 3. Simulation of a 1000 t oil spill off Zeebrugge, during calm weather conditions. Snapshots of the surface slick, as computed by MU–SLICKLETS, are shown 24 (a), 48 (b), 72 (c) and 96 (d) hours after the release. The solid line indicates the trajectory of the centre of mass computed by MU–SLICK.

2.3.1. The transport sub-model in MU-SLICK

MU-SLICK expects sea currents and surface winds at the nodes of a regular grid. In routine operations the currents are provided by the operational hydrodynamic model (MU-STORM) routinely operated at MUMM. The winds used are the same as those used in the hydrodynamic computations (*i.e.* meteorological forecasts by the UK Met. Office). The option also exists to use winds provided by the user (either constant or evolving with time).

Currents and winds are combined to give the velocity of the centre of mass of the oil slick:

$$\underline{u}_o = \underline{u}_w + \alpha \cdot \underline{\underline{D}} \cdot \underline{u}_a \qquad (1)$$

where

$\underline{u}_o$ is the velocity of the centre of the slick,

$\underline{u}_w$ is the depth-averaged current,

$\underline{u}_a$ is the wind velocity at 10 meters above the sea level,

α is the wind drift factor (3.15%),

$\underline{\underline{D}}$ is the transformation matrix $\begin{pmatrix} \cos\theta & \sin\theta \\ -\sin\theta & \cos\theta \end{pmatrix}$

θ is equal to $40° - 8\sqrt{u_a}$ when $0 \le u_a \le 25$ m/s and is set equal to zero for wind speeds greater than 25 m/s.

2.3.2. The spreading sub-model in MU-SLICK

The slick is assumed to be circular with a spatially (but time-varying) thickness. The oil is to spread under the influence of five forces: spreading forces are gravity and net surface tension, retarding forces are inertia, internal viscous force and interfacial (oil–water) viscous force. The acceleration of the edge of the oil slick is derived from the dynamic equilibrium of an elementary sector and reads:

$$\ddot{R} = \frac{3gV\Delta}{3\pi R^3} + \frac{3\pi\sigma R}{\rho_o V} + \frac{\dot{R}^2}{R} - 2.62817\frac{\rho_w\sqrt{\nu_w}}{\rho_o V}(R\dot{R})^{1.5} - 3\frac{\nu_o}{R^2}\dot{R} \qquad (2)$$

where

$\ddot{R}$ is the acceleration of the edge of the slick, $\dot{R}$ the spreading rate and R the radius of the slick,

g is the gravity,

V the total volume of the surface slick (oil & emulsion),

ρ_o, ν_o are the density and the cinematic viscosity of the oil,

ρ_w, ν_w are the density and the cinematic viscosity of the sea water,

$\Delta = \dfrac{\rho_w - \rho_o}{\rho_w}$,

$\sigma = \sigma_{wa} - \sigma_{oa} - \sigma_{ow}$ is the net surface tension.

The initial radius is usually set to $k\sqrt{V_0}$ (where k is a dimensional factor equal to 1 $m^{-1/2}$ and V_0 the initial volume of the spill) and the initial spreading rate is set to $0.65\sqrt{\Delta g}$. The model also gives an estimate of the thickness range, allowing 90% of the slick volume to cover 20% of the slick.

2.3.3. The weathering sub-model in MU-SLICK

The sparse information usually available when an accidental spill occurs limits the formulation of the weathering processes to relationships giving a good order of magnitude of the quantities involved, which can be done without the results of sophisticated oil analysis. Furthermore, since MU-SLICK is designed to provide information during the early stages of a spill, only the processes which could be significant on the short term – one or two weeks – are taken into account. The standard values of the constants are given in Table 2.

Evaporation

The evaporation rate is based on BLOKKER's formula[5] and reads

$$\dot{V}_e = \frac{\pi}{4} K_{ev} 2^{2-\beta} u_a^{\alpha} C_{14} R^{2-\beta} pM / 60 \tag{3}$$

where K_{ev} =1.2 10^{-8} (neutral atmosphere), $\beta = n/(2+n)$, $\alpha = (2-n)/(2+n)$, n is a turbulence parameter taken equal to 0.25, C_{14} is a coefficient that reflects the "skin" effect and pM is the product of the vapour pressure by the molecular weight. Note that only a fraction (ϕV_o) of the oil is allowed to evaporate.

Aerosols

$$\dot{V}_{sa} = C_6 V_r H \tag{4}$$

where V_r denotes the "untransformed" oil volume at the surface and H is a quantifier of the sea surface turbulence (related to the "significant wave height").

Vertical dispersion

$$\dot{V}_d = C_5 V_r H + C_{5chem} \tag{5}$$

where C_{5chem} accounts for the use of artificial dispersion means.

[5] BLOKKER, P.C. (1954), *Spreading and evaporation of petroleum products on water, Rapports du 4e Congrès Portuaire International*, Antwerp, Belgium.

Dissolution

$$\dot{V}_{\delta} = C_7 V_r \tag{6}$$

Mechanical recovery

$$\dot{V}_{mr} = C_{17} \tag{7}$$

Emulsification

$$\dot{V}_{em} = \frac{K_{em}}{C_{15}} V_r H - C_{17} \frac{V_{em}}{V_t} \tag{8}$$

$$\dot{V}_{w} = \frac{C_{18}}{1 - C_{18}} \frac{K_{em}}{C_{15}} V_r H - C_{17} \frac{V_{w}}{V_t} \tag{9}$$

where V_{em} is the volume of oil caught in the emulsion, K_{em} represents the ability of the oil to form emulsion, and C_{18} is the ratio of the water volume in the emulsion to the total volume of the emulsion.

Table 2. *Constants used in the weathering module of the MU-SLICK model.*

Constant	*Value*	*Constant*	*Value*
K_{ev}	$1.2\ 10^{-8}$	C_{14}	0.02
C_5	$3.0\ 10^{-6}$ $m^{-1}s^{-1}$	C_{15}	$5.0\ 10^{-7}$ m.s
C_{5chem}	user input	C_{17}	user input
C_6	$1.0\ 10^{-8}$ $m^{-1}s^{-1}$	C_{18}	0.8
C_7	$4.0\ 10^{-10}$ s^{-1}		

The physical properties of the oil evolve as follows:

Density
The apparent density of the slick depends on the densities of the evaporable and non-evaporable fractions of the oil and on the quantity of water present in the emulsion.

Surface tension

$$\sigma = \sigma(0)\frac{V_r}{V_r + V_{em}} \tag{10}$$

Viscosity

$$\nu_o = \nu_o(0)\left[\frac{V_r + V_{em}}{V_w + V_r + V_{em}}\right]^{-4} \tag{11}$$

2.3.4. The transport sub-model in MU–SLICKLETS
The drift of each individual slicklet results from the following equation:

$$\underline{u_o} = \underline{u_w} + \alpha' \cdot \underline{\underline{D}} \cdot \underline{u_a} + \underline{u_s} \tag{12}$$

where

$\underline{u_o}$ is the velocity of a given slicklet,

$\underline{u_w}$ is the depth-averaged current,

$\underline{u_a}$ is the wind velocity at 10 meters above the sea level,

α' is the wind drift factor, taken equal to $2.15 + 2 \cdot R[0:1]$,

$R[0:1]$ is a uniform pseudo-random function varying between 0 and 1,

$\underline{\underline{D}}$ is the transformation matrix $\begin{pmatrix} \cos\theta & \sin\theta \\ -\sin\theta & \cos\theta \end{pmatrix}$

θ is equal to $35° - 8\sqrt{u_a} + 10 \cdot R[0:1]$ when $0 \le u_a \le 25$ m/s and is set equal to $5° - 10° \cdot R[0:1]$ for wind speeds greater than 25 m/s,

$\underline{u_s}$ is the "spreading velocity" of the slicklet (its intensity varies from 0 to $\dot{R}$ and its direction from 0° to 360°, as a function of the initial relative position of the SLICKLET).

At the beginning of the simulation, the slicklets are evenly distributed in space within the circle given by the initial conditions of MU-SLICK. After each iteration the position of the slicklets is checked against a given bathymetry and they are possibly flagged as "beached".

2.3.5. The sub-surface drift sub-model in MU–SLICKLETS

"Large" oil droplets entering the water column under the influence of the surface turbulence (waves) rise to the surface later due to their buoyancy, accounting for an additional "spreading" effect: their horizontal velocity under the surface is different from that of the surface slick.

This process is modelled by randomly selecting slicklets at each time step and having them *move* to a given depth under the surface. The selection algorithm and the depth are computed according to a pseudo–random technique in a range depending on the significant wave height *H*.

The particle resurfaces with a velocity that is determined by the buoyancy of the oil (Δ). During the time needed to come back to the surface again, it drifts with a horizontal velocity:

$$\underline{u}_o = \underline{u}_w + \alpha'' \cdot \underline{\underline{D}} \cdot \underline{u}_a \tag{13}$$

where
α'' is a wind drift factor, varying linearly between $2.15 + 2 \cdot R[0:1]$ at the surface and 0 at a reference depth depending on the wind intensity u_a.

2.3.6. The weathering sub-model in MU–SLICKLETS

At a given time step, all the slicklets represent the same volume of oil, which is computed as the total volume of the reference slick at that time divided by the number of "active" slicklets (*i.e.* those which are not flagged as "beached" or which did not reach the limits of the computational domain). In order to preserve mass conservation, there is a feedback mechanism reflecting the beaching of slicklets on the volume of the reference slick.

2.4. Fate in the water column

Classical advection–diffusion models can be used to track the oil that enters the water column. It is known[6], however, that such a model is rather difficult to test and to tune.

---

[6] PROCTOR, R. (ED), BAART, A., BERG, P., BOON, J., DELEERSNIJDER, E., DELHEZ, E., GARREAU, P., GERRITSEN, H., JONES, J.E., DE KOK, J., LAZURE, P., LUYTEN, P., OZER, J., POLHMANN, T., RUDDICK, K., SALDEN, R., SALOMON, J.C., SKOGEN, M., TARTINVILLE, B.

The type of effect and impact functions made available in the frame of the project estimate the effects of the presence of a pollutant in the water column and on the bottom in a rather integrated manner. We therefore decided to take advantage, for estimating the concentrations levels in the water column on the one hand, of the "fine" – in time and space – knowledge of the quantities that enters the water body provided by the particles approach and, for estimating the deposition on the bottom on the other hand, to extend the random advection techniques to represent the sedimentation process.

2.5. Fate in the atmosphere

The surface behaviour model computes the flux of pollutant to the atmosphere. These values, defined in time and space, are available as input to an atmospheric dispersion model. The MU–ATMOS model[7], for instance, is designed to be coupled to such a surface behaviour model.

This is a so–called "puff" model, where releases to the atmosphere are considered as successive independent input sources of a given quantity of pollutant ("puffs") that evolve independently (drift and Gaussian dispersion). The concentrations corresponding to these puffs are then linearly combined to give a first order estimate of the concentration field of the pollutant in the atmosphere.

Unfortunately, in the frame of the MARE–DASM project, no effect or impact functions of possibly harmful concentrations in the atmosphere could be made available. The effective coupling of the atmospheric dispersion model to the surface dispersion model has therefore not been implemented.

2.6. Global repartition in the environment

MACKAY *et al.*[8] define four levels of models for describing the repartition of a pollutant in the environment. "Level I" models have no dynamics included and assume an instantaneous repartition of the pollutant between the various compartments, under steady-state conditions. The higher the level, the higher the complexity of the processes modelled and, hence, the more detailed the information

---

& VESTED, H.J. (1997), *NOMADS – North Sea Model Advection-Dispersion Study (MAS2-CT94-0105). Final Report,* POL Internal Document No. 108; PROCTOR, R., (ED), DAMM, P., DELHEZ, E., DUMAS, F., GERRITSEN, H., DE GOEDE, E., JONES, J.E., DE KOK, J., OZER, J., POLHMANN, T., RASCH, P., SKOGEN, M. & SORENSEN, J.T. (2002), *North Sea Model Advection-Dispersion Study 2: Assessments of model variability (MAS3-CT98-0163). Final Report,* POL Internal Document No. 144.

[7] DEMUTH, Cl. & SCORY, S. (1986), *ATMOS : Système de simulation du comportement des substances flottantes et volatiles déversées en mer*, Commission des Communautés Européennes, 6621/85850014. (in French)

[8] MACKAY, D., DI GUARDO, A., PATERSON, S., KISCI, G. & COWAN, C.E. (1996a), Assessing the fate of new and existing chemicals: a five stage process, *Environmental Toxicology and Chemistry*, 15, 1.618-1.626.

needed to run the model (characteristics of the substance, parameters governing the processes, *etc.*).

According to MACKAY's classification, the surface behaviour and the water column modules described above form a "Level II" model for oil considered as a homogeneous product. But oil is a mixture of hundreds of chemical compounds. For some purposes, considering this mixture as a single product yields enough information to estimate the effects of the pollution, but not for all.

Estimating the effects of the pollution on living resources, for instance, implies having a closer view of the fate of oil components prone to induce toxic effects.

Due to the lack of relevant data to apply the MACKAY model to the toxic components, VANDENBROELE *et al*[9]., in their case study related to oil pollution, apply an *a priori* factor (80% of the initial water-soluble fraction of the oil) to estimate the concentration in the – fully mixed – water column. For each of the identified toxic compounds, an average factor found in the literature is then used to estimate the specific concentration of that compound.

HEIJERICK & JANSEN[10] applied two "Level I" models for a variety of chemical compounds, using two specific reference environment: a "*standard marine environment*" and a "*reference marine environment*" where the dimensions of the compartments have been chosen to correspond to the characteristic dimensions of an area affected by an accidental oil pollution incident.

These model runs provide a synthetic view of the relative repartition of key chemicals in standardized environments. They show that the initial quantity of a substance released in the environment has almost no influence, according to these models, on the relative distribution amongst the various compartments. The relative dimensions ("volumes") of the various compartments with respect to each other, however, sensibly influence the repartition of the pollutant amongst these compartments. HEIJERICK & JANSEN conclude that their results can be considered as giving the general trends of the repartition of a given substance in the compartments of the environment.

It is thus clear that the present state of the art in identifying the individual behaviour of each of the toxic compounds to be found in "oil" lacks precision, mainly due to the lack on information on the precise composition of oils and on the specific behaviour of these compounds in the marine environment.

---

[9] VANDENBROELE, M., VANGHELUWE, M., JANSSEN, C., PERSOONE, G. & LE RCY, D. (1997), Activiteitenverslag. Impulsprogramma Zeewetenschappen: Definiëring en toepassing van ecologische criteria en economische indicatoren voor de effectstudie en kostenbepaling van diverse types van verontreiniging in de Noordzee.

[10] HEIJERICK, D. & JANSEN, C. (2002), Identificatie en kwantificatie van de verschillende factoren die bijdragen tot het degradatie van het mariene milieu, *in: MARE–DASM project, Final report.*

In our implementation, practically, at each point of a 500m×500m grid the quantity of oil that entered the water column (expressed in terms of volumes) provided by the surface behaviour model is converted into a mass (taking into account the evolution of the pollutant density over time) and a concentration (taking into account the local depth).

2.7. Effects

Oil at sea has a potentially adverse effect on all the marine resources, living or non-living, including the human activities making use of these resources.

For the purpose of this study, these possible effects are split into two categories, depending most of the time on the living or non-living nature of the considered "target":

(i) effects due to the presence of the pollutant only;
(ii) effects where the characteristics of the pollutant, its toxicity principally, are important.

The basic results of the surface model (volume of oil at the surface – or on the beach – as a function of time and location) are sufficient to estimate the effects belonging to the first category. Amongst these one can cite:

(i) birds, as the most significant impact occurs when they are in direct contact with the slick;
(ii) pollution of the coastline;
(iii) perturbation of human activities;
(iv) …

To estimate the effects on these targets, simple logical and arithmetic functions are used. For a given simultaneous presence of oil and a target, the relevant metric dimension of the oil slick (volume, area, thickness or a combination thereof) is tested against reference minimum and maximum values, corresponding to no or maximum effect respectively. The assumption is made that variation occurs linearly between these two values. The total effect for a given target is then obtained by multiplying the effect by the quantitative characteristics of the target and by integrating the resulting values over the concerned area or coastline length.

2.7.1. The Biological Effects Sub-Model (BESM)

VANDENBROELE *et al*[11]. performed an extensive study to define ecological criteria and economic indicators able to describe the effects of various types of

[11] VANDENBROELE, M., VANGHELUWE, M., JANSSEN, C., PERSOONE, G. & LE ROY, D. (1997), *o.c.*

pollution. A "Biological Effects Submodel" (BESM) was developed in order to describe the short-term consequences of a pollution event on the biota in terms of direct mortality and indirect losses due to cumulative effects through the food chain.

In order to take into account the spatial heterogeneity of the populations considered, the Belgian Continental Plate (BCP) was divided into four ecological zones where the living resources are assumed to have homogeneous characteristics. These zones are depicted on Figure **4**.

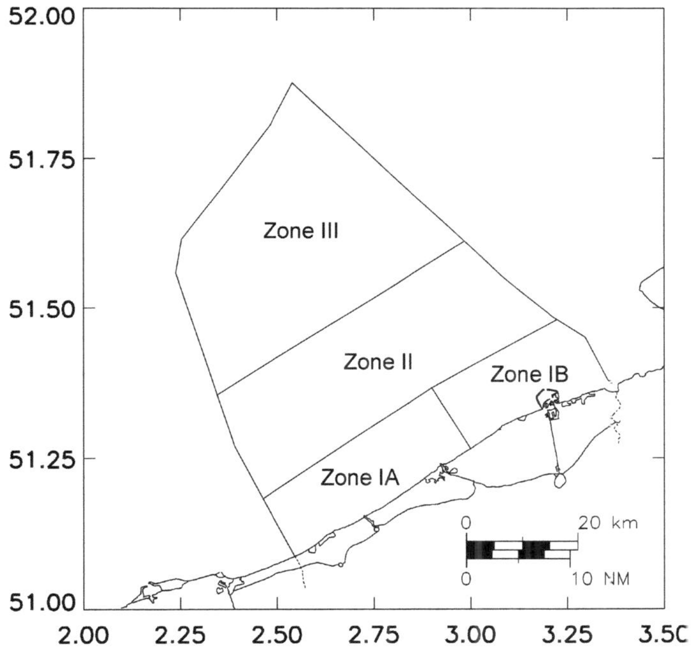

Figure 4. *Ecological zones of the Belgian continental plate as defined in VANDENBROELE et al.*

There are seven target populations ("compartments" of the ecological model): phytoplankton, zooplankton, benthos, epibenthos, demersal fishes, pelagic fishes and birds.

There are many uncertainties about the combined effects of the toxic compounds present in "oil". Therefore, a first assessment of the direct impact of the toxic compounds known to be present is made for each of the compartments and the compound that induces the highest impact is retained for further quantification.

Direct losses

The direct mortality due to the interaction between a pollutant and a population is computed according to a log–normal function of the concentration:

$$p_0 = \frac{1}{\sqrt{2\pi}} \int_{-\infty}^{X_0} e^{-\frac{X^2}{2}} dX \tag{14}$$

where:

p_0 represents the fraction of the population that will die when exposed to a concentration C_0,

C_0 is the concentration in the water column,

$X = \dfrac{\log(C) - \mu}{\sigma}$,

$\mu = \log(LC_{50})$, where LC_{50} is a value found in the literature, corrected for possible differences between reference temperature and exposure time,

σ, as a first approximation and in conformity with data found in the literature for hydrocarbons, is set equal to 0.83.

The formula for estimating the direct loss in a given compartment i is then written as:

$$\Delta P_i = 12.5 \times p_{0i} \times P_i \times \Omega \times \Delta t \tag{15}$$

where:

p_{0i} represents the fraction of the population of compartment i that will die when exposed to a concentration C_0,

P_i is the normal production of compartment i (see below) [gC/m²]

Ω is the surface of the polluted area,

Δt is the exposure time.

In the MARE–DASM system, the BESM module processes the outputs of the surface slick model. The fraction of oil that entered the water column at a given place is supposed to mix over the full depth. This first–order assumption is valid for the zone considered. According to our existing knowledge on the behaviour of the toxic compounds in the marine environment and available data on their toxicity, a first "guess" of the impact is computed in terms of the direct biomass losses, assuming the biological parameters are spatially homogeneous. Further in the processing, a finer estimate is given by taking into account the spatial variation of these parameters.

Indirect losses

A stationary ecological model quantifies the fluxes between the seven trophic levels. A basic sketch is given in Figure 5.
The direct losses found above are propagated through that system, assuming a constant assimilation efficiency, e:

$$\Delta P_i = e \times a_{i-1,i} \times \Delta P_{i-1} \tag{16}$$

with e set equal to 0.2. The total primary production ($i = 1$) of the Belgian continental plate is estimated as 410 gC/m² on yearly basis, or 1.123 gC/m² a day. The P_i at the higher levels are inferred from those values. The $a_{i-1,i}$ are those given in Figure 5.

The direct losses computed before are propagated in the ecological system according to the formulae given above. The resulting figures are then summed for each of the compartments. A first "guess" of the impact in terms of the indirect biomass losses is computed, assuming the biological parameters are spatially homogeneous. Further in the processing, a finer estimate is given by taking into account the spatial variation of these parameters.

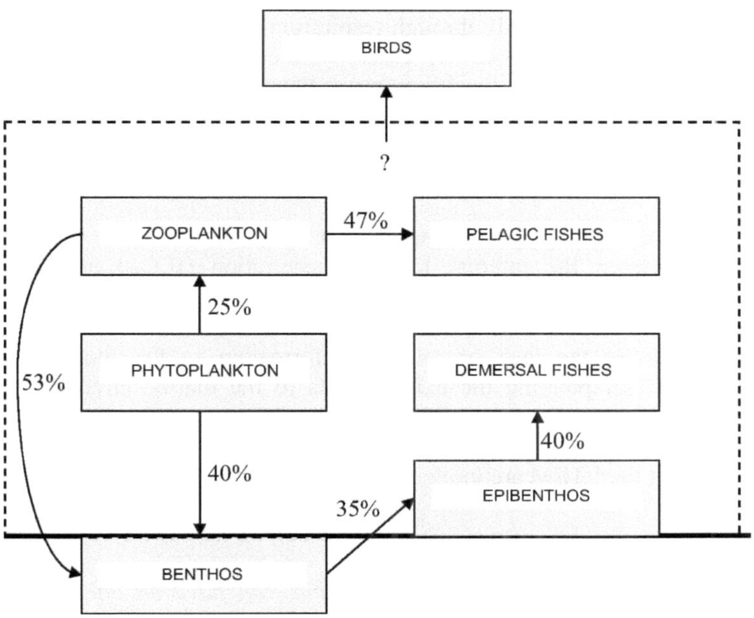

2.7.2. Seabirds

According to VANDENBROELE *et al.*, neither the oil dissolved in the water nor the uptake of polluted food significantly impact seabird populations. The main effect results from direct contact between the birds and the oil (at the surface of the sea).

The direct loss in bird populations is thus computed according to:

$$\Delta B_i = S_i \times r_i \times \Omega' \times B_i \tag{17}$$

where

S_i is the sensitivity index of species i,

B_i is the population density of species i,

r_i is a factor for converting sensitivity into mortality, taken equal to 0.031,

Ω' is the surface of the polluted area (Ω) extended by a 5 km–wide buffer zone.

2.7.3. The Long-Term Effects Model (LTEM)

HEIJERICK & JANSEN[12] concentrate on the prediction of long-term effects. Due to the lack of relevant data, they build their model ("LTEM") on the "worst case" edge of the possible assumptions. The model includes a two-step calculation. In a first step the uptake of pollutant by the organisms is computed. Uptake occurs either directly from the water (mainly through respiration) or via the food. The food web scheme underlying this process is basically the same as that in the BESM. HEIJERICK & JANSEN significantly improve the way contaminants are transferred from one level to the other by taking into account the dynamics of the system: assimilation efficiency, duration of the exposure to the contaminant, characteristic (re)generation time of the populations constituting a given level.

In a second step, the LTEM computes the long-term effects, using a linear relationship between the internal lethal concentration (ILC_{50}) and the natural logarithm of the exposure time. As is often the case, a more detailed theoretical model requests more information than a simpler one for effective use. The authors several times stress the lack of relevant information in the literature and the uncertainties in extrapolating the existing data to the marine environment. Total losses computed by means of the LTEM must thus be taken with extreme care.

The equations of the LTEM are implemented as such in the MARE–DASM system.

Note on the effect of dispersants:

HEIJERICK & JANSEN state that "*dispersants can have an effect on the bio–availability and the toxicity of oil components. The present knowledge thereabout*

[12] HEIJERICK, D. & JANSEN, C. (2002), *o.c.*

and the very scarce information on the specific effects of dispersants (...) do not allow to take their effects on the toxicity into account in the assessment of long–term effects" (free translation). Use of dispersants will therefore only be reflected in the model by increasing the value of the parameters governing the (vertical) dispersion and the dissolution of oil in the water column.

2.8. Impacts, valuation and costs

Finally, *assessing the consequences of accidental spillage of oil at the environmental and socio-economic levels* implies translating, wherever possible, the effects of the pollution on the natural goods quantified in the preceding steps into monetary values and identifing and evaluating the direct costs and economic losses generated by the pollution itself.

The costs identified in VANDENBROELE *et al.*[13] are implemented as such in the MARE–DASM system of models (and the underlying economic information updated wherever possible):

(i) costs generated by the pollution event (monitoring, co-ordination, ...), which can be considered as independent of the extent of the pollution above a threshold extent;

(ii) costs generated by the pollution event and proportional to its extent (combating actions at sea, cleaning of the beaches, ...);

(iii) commercial losses (losses in fish stocks induce a decrease of the fishery activity, the pollution of the beaches induces a decrease in the tourist activity, ...) derived from information produced by surface behaviour model (polluted coastline) and the "effects" sub–models.

In the frame of the present study VAN BIERVLIET *et al.*[14] concentrated on the calculation of the restoration costs and, mainly, on the valuation of natural goods that commonly do not have a market price. The restoration costs considered only address the expenses associated with the revalidation and breeding (farming) of animals belonging to the higher trophic levels (fishes, marine mammals, birds). No figure is given for the possible restoration of habitats. In the integrated system of models these revalidation and breeding/farming costs (given per individual) are applied to, respectively, a percentage of the population of the considered species hit by the pollution at sublethal concentrations and to the total losses for that population, as given by the "effects" modules.

---

[13] VANDENBROELE, M., VANGHELUWE, M., JANSSEN, C., PERSOONE, G. & LE ROY, D. (1997), *o.c.*

[14] VAN BIERVLIET, K., BOGAERT, G., DECONINCK, M., LE ROY, D. & BOGAERT, S. (2002), Ontwikkelen van socio-economische beoordelingscriteria die het mogelijk maken de kostprijs van deze degradatie objectief te bepalen., *in: "MARE–DASM project, Final report"*. See also *supra* in this book VAN BIERVLIET, K., LE ROY, D. & NUNES, P.AL.D., *A contingent valuation study of an accidental oil spill along the belgian coast* (Part III, Chapter 8).

In order to be able to evaluate the natural goods that do not have a commercial ("usage") value, VAN BIERVLIET *et al.* performed an extensive study using the so-called "Contingent Valuation Method". Briefly, this method uses the concepts of "Willingness to pay" (to avoid a pollution) or "Willingness to accept" (the consequences of a pollution) expressed by a reference population during a survey to identify the "price" these natural goods would have if included in the commercial market. From the results of their survey, the authors conclude that the "price" so defined of the Belgian part of the North Sea would amount between 375 and 606 million euros. But they also note that this "price" is not linked to the scale of the pollution that could affect the marine environment. In other words, no function could be found between an impact and a price. Evaluation of non–commercial natural goods is therefore not included in the integrated model.

2.8.1. Impact at sea

The impact at sea is due to the presence of oil on the sea surface, in the water column and on the sea bottom.

The effects on the biota were already described. Given the presently existing human activities at sea no significant impact on other activities than fishery is expected due to the presence of oil at the surface.

2.8.2. Impact at the coast

The coastline is divided into segments of arbitrary length (500m in our application). Each of these segments is geo-referenced and has a number of "attributes":

(i) length;
(ii) type (sandy, sandy with coastal defences, rocky, dam, estuary, …)
(iii) ecological value;
(iv) existence of human activities (recreational stay, bathing waters, supply water, jetsky, sailing, recreational fishery, commercial fishery, fishery harbor, ferry harbor, "cargo" harbor, …);
(v) specific status (protected area, …);
(vi) specific details (*i.e.* relevant elements that have a shorter characteristic length than the segment, see below);
(vii) ...

Each segment belongs to a community, which, in turn, also has a number of specific attributes:

(i) limits;
(ii) number of inhabitants split over four major categories:
  a. recreational stays:
    i. one-day stays ("A");
    ii. stays during more than one day (*i.e.* people hiring a house or a flat; "B1" from two days to a week, "B2" from one

week up to one month, "B3" more than one month);
iii. semi-permanent stays (*i.e.* people owning their own house or flat and using it as a "second residence"; "C1" from two days to a week, "C2" from one week up to one month, "C3" more than one month);

b. permanent inhabitants (category "D").

Provision has been made to take into account details that have a characteristic length shorter than the reference segment length, *e. g.* harbor entrances, some small estuaries (*'t Zwin*), ... Harbors, for instance, might also have a number of specific attributes:

(i) tons/day commercial fishes (2);
(ii) number of ferry movements/day;
(iii) number of tons/day cargo.

According to the characteristics of the coastline segments hit by the pollution, a number of quantitative and qualitative outputs are generated:

(i) total volume of stranded oil, maximum density of oil on the beach (m^3/m);
(ii) length of the coastline of a given ecological value;
(iii) ...

2.8.3. Cost function at sea

The losses in biomass computed according to the formulae given in Section 2.7 are converted into commercial losses by multiplying them by the proportion of the commercial species (in each of the relevant compartments: crustaceans, demersal fishes, ...) and by the current price at landing.

2.8.4. Cost functions at the coast

As mentioned in VANDENBROELE *et al.*[15] , costs of cleaning-up the beaches vary as a function of:

(i) the location (accessibility of the polluted zone and type of coastline);
(ii) the type of oil (the more viscous, the more onerous to clean);
(iii) the quantity of oil (but they state there is no clear relationship between quantity and costs);
(iv) the technique(s) used to clean up the coast;
(v) the cost of waste treatment, taking into account the composition of the wastes (*e.g.* oil/sand mixture and the corresponding relative proportions), the distance to the waste treatment plant and the costs of the treatment itself.

---

[15] VANDENBROELE, M., VANGHELUWE, M., JANSSEN, C., PERSOONE, G. & LE ROY, D. (1997), *o.c.*

The authors do not draw explicit relationships between a metric of the pollution level and the various elements of the total cost listed above. They list figures found in the literature, showing indeed a great variability in clean-up costs, hence exemplifying their multi-variate nature.

It would have been possible to write down a "formula" taking into account all these parameters. It would, however, be of no practical use most of the time. The form of the implemented function provides the user a way to summarize his/her knowledge of the basic costs into three parameters:

$$\begin{aligned} &if\ SD \le SD_t,\ C_{cleanup} = C_1 \\ &if\ SD > SD_t,\ C_{cleanup} = C_1 + C_2 \times (SD - SD_t) \end{aligned} \tag{18}$$

The parameters are defined in Table 3.

Table 3. *Parameters characterizing the costs bound to coastline cleanup.*

Variable	*Description*	*Unit*	*Default value*
SD	stranded density	m^3 oil/m coastline	N/A
SD_t	threshold value	m^3 oil/m coastline	0.00005
$C_{cleanup}$	cost of cleaning up the beach	€	N/A
C_1	fixed cost	€/m coastline	1.00
C_2	cost tied to pollution level	€/m^3 oil	2000.00

2.8.5. Valuation of impact on tourism

VANDENBROELE *et al.* express the loss of income due to the impact on tourism of pollution reaching the coastline as a function of:

(i) the importance of the pollution (will the shore be closed to tourists or not);
(ii) the resort;
(iii) the number of recreational stays (split into seven categories and sub-categories, see the attributes defined for a "community" above) split over the months of the year.

The first parameter is more a result of a political decision than the result of an unambiguous and objective function of the, *e.g.*, quantity of oil that hit the coast. It may also depend on the importance of the human and technical means mobilised to clean up the beaches and on the efficiency of their work, what cannot easily be parameterized. We therefore decided to implement this function in such a way that both options are always computed.

The original general formula for computing the global loss of income is given by:

$$IL_{tourism} = \int_{t_s}^{t_e} e^{-rt} E_{tourism} \sum_j [S_j^0(t) - S_j^p(t)]\, dt \tag{19}$$

where:
$IL_{tourism}$ is the income loss due to the pollution,
t_s is the starting time of the event (Note: the authors define it as the initial time of the release *–tijdstip van de lozing–*; we define it as the time when the oil begins to reach the coast),
t_e is the time when the beach is back to its normal state,
$E_{tourism}$ is the expenses made by the tourists according to the various categories of stays,
j is the index pointing to the various communities,
S_0 number of "stays" according to the various categories, in normal conditions,
S_p number of "stays" according to the various categories, when the beach is polluted,
r is a decay coefficent, and
t is the time.

This formula thus contains an implicit summation on the type of stays. We therefore prefer to re-organize it in a way that a finer description can be given, if the corresponding data are available:

$$IL_{tourism} = \sum_j \sum_k \int_{t_s}^{t_e} e^{-rt} E_{tourism(j,k)} [S_{j,k}^0(t) - S_{j,k}^p(t)]\, dt \tag{20}$$

where k is an index pointing to the various categories of recreational stays,

Some simplification of formula (20) is possible by averaging at the monthly level and taking into account that $S_k^0 - S_k^p = NC_k\, S_k^0, 0 \le NC_k \le 1$:

$$IL_{tourism}(m) = \sum_j \sum_k E_{tourism(j,k)}\, F_{j,k,m}\, S_{j,k}^0\, NC_k \tag{21}$$

where:
m refers to the month of the year,
NC_k is set to 1 when the coast is supposed to be closed, and

$F_{j,k,m} = F_j \times F_k \times F_m$ gives, as a percentage, the number of stays in the various categories, according to the month of the year and the resort.
E and S are figures that might be found, for instance, in statistics regularly issued by the tourist federations. In order to estimate F and NC_k, the authors performed a field inquiry. The values given below and in the following tables, unless otherwise noted, are inferred from tables of results given in PERSOONE *et al.*.

$E_{tourism}$, total income for the Belgian coast thanks to tourism: 830.44 to 892.42 M€. Taking into account the evolution of the cost of living, these amounts extrapolated to February 2003 (indice=112.32, base 1996=100) yield 932.75 and 1002.36 M€, respectively. The default value in our implementation is taken equal to 1000 M€.

Table 4. *Monthly variation of the total income (F_m).*

Month	1	2	3	4	5	6
Proportion	0.90 %	1.40 %	3.90 %	8.60 %	8.50 %	11.10 %
Month	7	8	9	10	11	12
Proportion	26.20 %	26.30 %	9.30 %	1.60 %	1.10 %	1.10 %

Table 5. *Repartition of the incomes according to the categories of stay, to be used in case of "large" pollution (F_k).*

Category	*Proportion*
A	27.00 %
B	50.50 %
C	22.50 %
	100.00 %

Table 6. *Repartition of the incomes according to the categories of stay, to be used in case of "small" pollution (F_k).*

Category	*Proportion*	*Sub-category*	*Proportion*
A	34.07 %		
B	45.58 %	B1	48.54 %
		B2	46.72 %
		B3	4.74 %
C	20.35 %	C1	48.55 %
		C2	46.72 %
		C3	4.73 %
	100.00 %		

Note: the original figures show slightly diverging percentages for B_3 and C_3 in April, with respect to the other months of the year. This might be the result of slightly different behaviour of the tourists during the Easter holidays. The difference is not significant, however.

Table 7. *Repartition of the incomes according to the communities (F_j).*

Community	*Proportion*
De Panne	6.20 %
Koksijde	14.80 %
Nieuwpoort	8.80 %
Middelkerke	15.90 %
Ostend	6.10 %
De Haan	26.50 %
Blankenberge	4.80 %
Knokke	16.90 %

And, finally, NC_k, the percentage of loss in case of a small pollution, is taken equal to 7.85 %.

Practically, the computation of these costs has been implemented in the following way.

(i) per community, the number of segments of its coastline where the density of the beached oil is greater than SD_t amounts to at least one third of the total number of segments constituting the coastline of that community, a loss of incomes bound to tourism is assumed;

(ii) the upper limit is computed according to:

a) $IL_{gl} = E_{tourism} \times F_m$ (first option, "large pollution", beaches are closed) and to

b) $IL_{gs} = E_{tourism} \times F_m \times NC_k$ (second option, "small pollution", beaches are not closed)

The impact on the major categories of stay can then be computed by multiplying the two results above by the values of F_k given in Tables 5 and Table 6 respectively.

The season of the year during which the event occurs is taken into account by the factor F_m (Table 3).

(iii) a more realistic estimate is computed by summing the partial losses, $IL_l(j) = E_{tourism} \times F_m \times F_j$ and $IL_s(j) = E_{tourism} \times F_m \times NC \times F_j$, respectively, for each of the communities identified at step 1. A more detailed impact on the various categories of stay is then computed the same way as is described at step 2.

2.9. Integration and practical implementation

As a result of the specific aspects of each of the modules described above, the initial conceptual scheme of the system (Figure 1) has been adapted as shown on Figure 6.

The use of two possible methods for estimating the repartition of the pollution in the environment and of two models for quantifying its effects on the living resources imply that several differing values for the same item may appear in the final "bill". This cannot be avoided and should be understood as the result of the many remaining uncertainties.

Regarding the technical aspects of the integrated system, it should be mentioned that the models are either written in FORTRAN (running on Unix systems) or are implemented as MS–Excel spreadsheets (running on MS–Windows system). In order to keep the implementation as efficient as possible, it has been decided to keep the most time-consuming models and the pure mathematical relationships running on Unix systems and to develop the geographically-based functions within a GIS environment (ArcView). For the user these technical details are "masked" by a unique (web-based) interface.

The MARE–DASM system of models is a chain of modules that must be run sequentially. The user interacts with the system to:

(i) define the parameters (input);

(ii) visualise the results (output).

The parameters have been sorted into three categories:

(i) the variables, *i.e.* the parameters which are specific to the case being studied (*e.g.* time and position of the spill, type of oil, ...);

(ii) the auxiliary parameters, *i.e.* the parameters that might be kept as constant for the purpose of the study (*e.g.* the various unitary costs);

(iii) the model parameters, *i.e.* the parameters incorporated in the functions solved in the model (*e.g.* the dispersion rate).

The user is allowed to modify all these parameters. It is however, strongly recommended that parameters belonging to categories 2 and 3 be handled with extreme caution in order to avoid totally unrealistic results or, even, numerical problems.

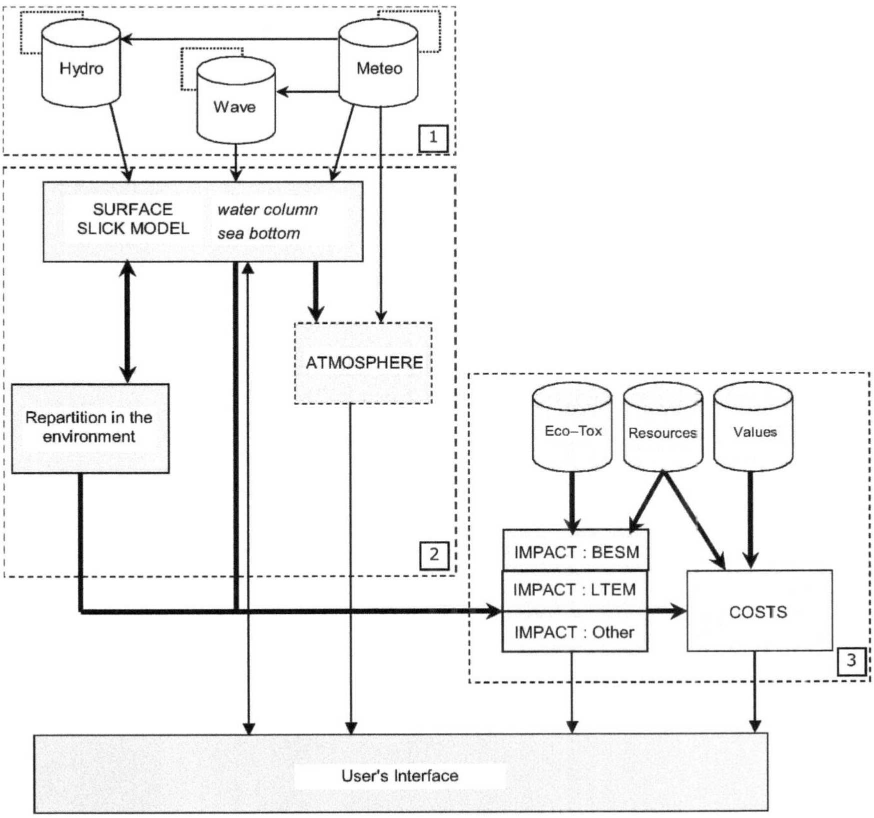

Figure 6*. Effective conceptual scheme of the integrated MARE–DASM system of models. 1: Environmental conditions; 2: Physical-chemical behaviour of the pollutant; 3: Effects, impacts and quantification.*

3. CASE STUDY

3.1. Description

The case study briefly described here corresponds to an hypothetical scenario based on recent incidents, namely the *Vera/Music* case (13 April 2001), the stranding of the *Voltaire* (11 December 2002) on a sand bank north of the "Vlakte van de Raan" and the *Tricolor/Vicky* case (1 January 2003).

The scenario assumes that a tanker, due to a collision or stranding, spills the full content of one tank, containing 500 m^3 of heavy fuel.

A few days after the *Vera/Music* release (estimated as 20 tons heavy fuel) oil beached on the Belgian coastline, leaving oil residues on a zone from 1.5 to 3 m wide along the high water line. The coverage of this polluted zone was estimated as 40 %.

The environmental conditions used in this case study (winds and currents) are those prevailing during the 2nd and the 3rd weeks of April 2001. The release is supposed to have occurred at position (51° 29' 47"N, 3° 01' 05"E) on the 13th at 6 pm (UTC).

3.2. Outputs of the spill model

The following section gives a digest of the many possible outputs of the model.

The maximum extent of the slick when hitting the coast (MU-SLICK diameter) is equal to 3.5 km and that event occurs on 15 April at 21:00 UTC.

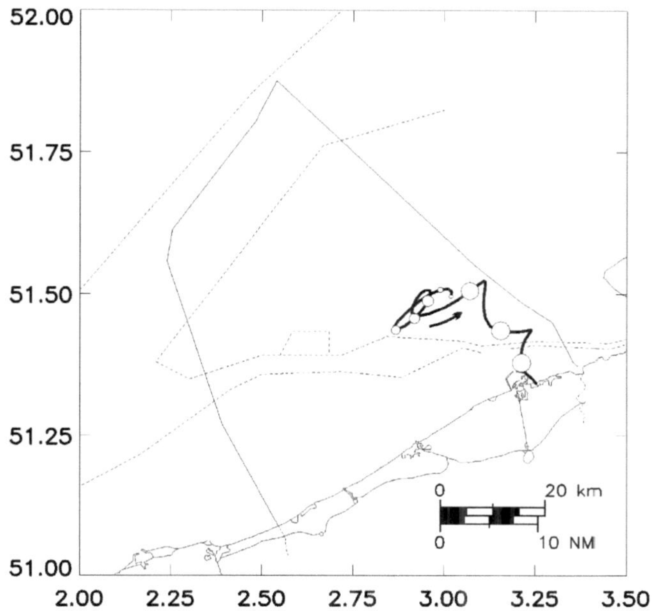

Figure 7: *Trajectory and spreading computed by the MU-SLICK module. For clarity slick extent is given at selected moments only.*

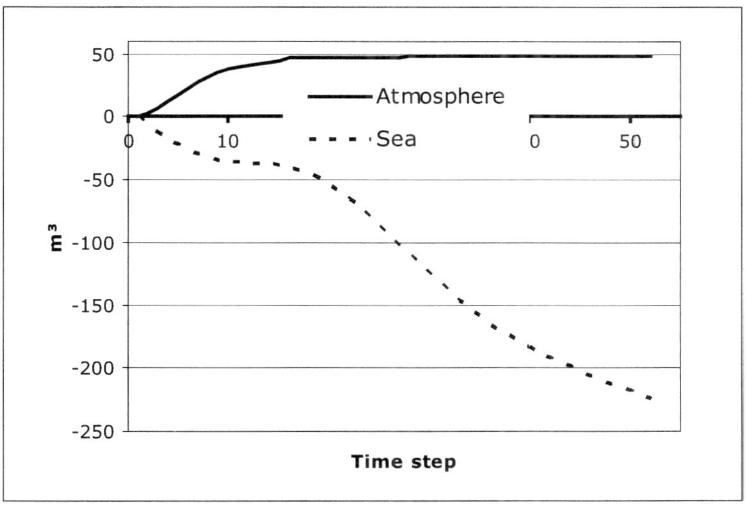

Figure 8. *Evolution over time of the oil volumes entering the atmosphere and the water column.*

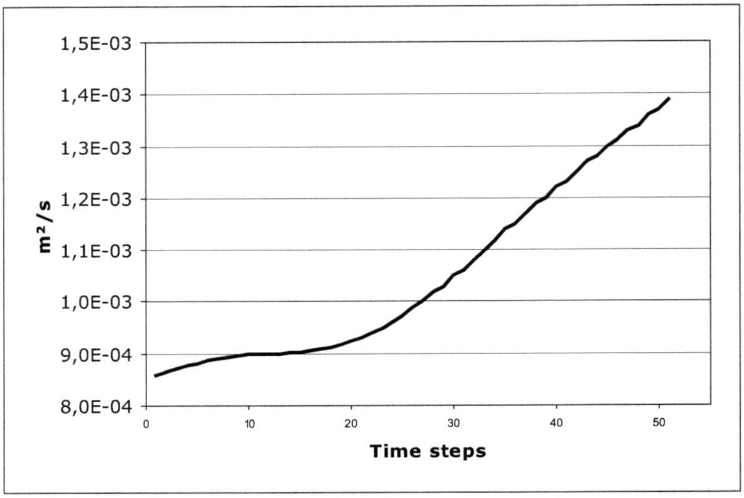

Figure 9. *Evolution over time of apparent kinematic viscosity of the oil and emulsion mixture.*

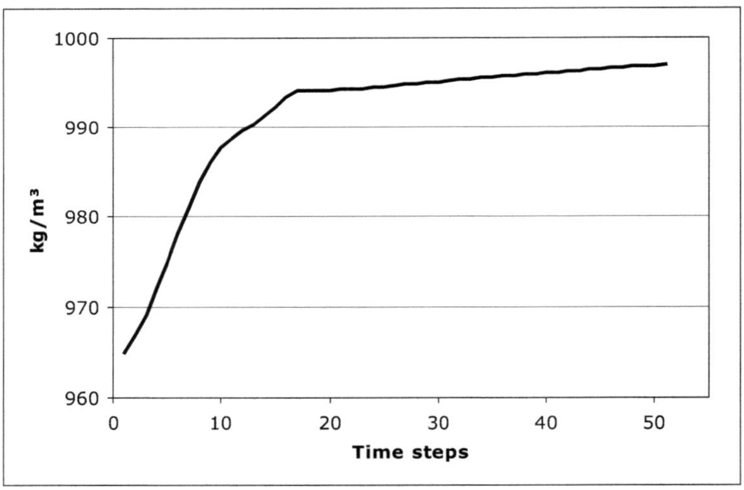

Figure 10. *Evolution of the apparent density of the oil and emulsion mixture.*

Table 8. *Final balance as computed by the physical behaviour models:*

	Oil in the atmosphere	*Oil in the sea*	*Oil on the sea/on land*	*Oil on the sea*	*Oil on the beach*
MU-SLICK	48.10 m³	221.10 m³	230.20 m³	–	–
	9.63 %	44.27 %	46.10 %		
MU-SLICKLETS	48.16 m³	230.31 m³	–	0.00 m³	221.52 m³
	9.63 %	46.06 %		0.00 %	44.30 %

The slight difference in the balance between the oil left at the sea surface (or beached) and the oil in the water column as given by the two sub-models is due to the different functions used to estimate the flux of oil entering the water column. The characteristics of the oil may, to some extent, increase or decrease that difference.

3.3. Physical impact on the coastline

The polluted coastline is shown in Figure 12.

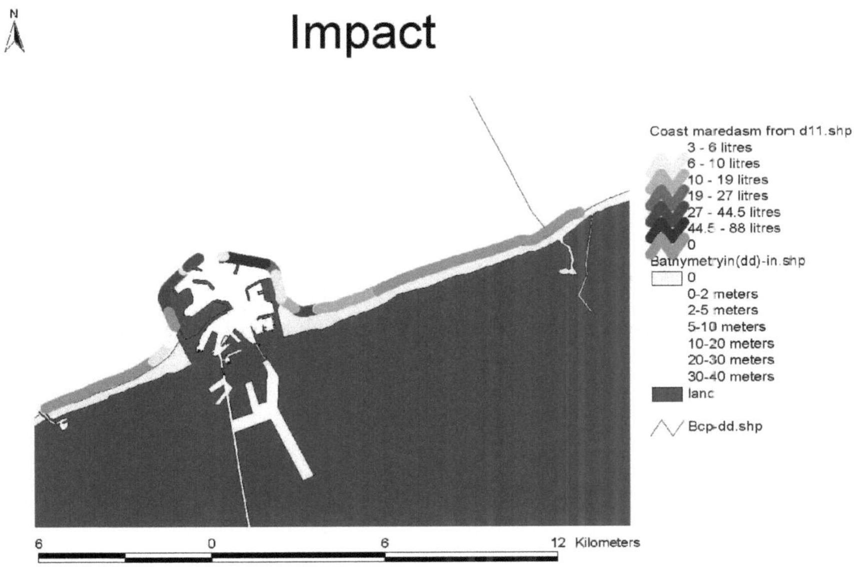

Figure 11: *Integrated system: zoom on the incorporation of the model outputs into the GIS environment.*

The following two tables summarize the physical impact on the coast. "Lengths" are given in meters of coastline.

Table 9. *Impact on the coastline according to type.*

Community	*m³ oil*	*Polluted length*	*Length polluted sandy beach*	*Length polluted beach with coastal defences*	*Length polluted port infrastructure*	*Length polluted port entrance*
Knokke Heist	43.4	4000	1000	3000	0	0
Zeebrugge	163.4	8000	1000	0	6500	500
Other	0.0	0	0	0	0	0
TOTAL	206.8	12000	2000	3000	6500	500

Table 10. *Impact on the natural coastline according to its legal protection status.*

Community	*Length polluted zone under Birds Directive*	*Length polluted zone under Habitats Directive*	*Length polluted zone under Dune Decree*	*Length polluted regional nature preservation zone*
Knokke Heist	0	1000	1000	1000
Zeebrugge	500	0	0	0
Other	0	0	0	0
TOTAL	500	1000	1000	1000

3.4. Physical impact at sea

Figure 12 gives an overview of the trajectory of the slick and of its extent at selected moments.

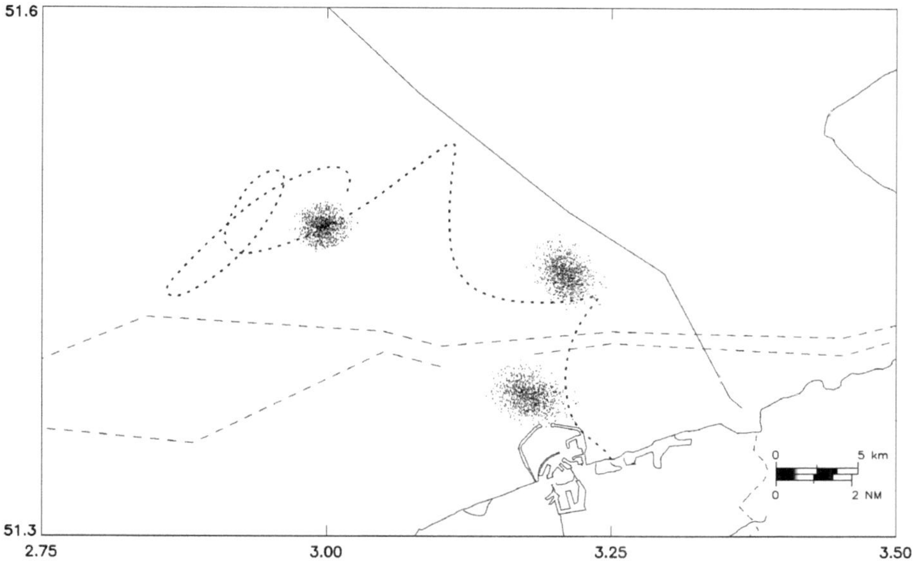

Figure 12. *Zoom on the zone affected by the pollution, showing the global trajectory as computed by MU–SLICK and some instantaneous snapshots of the spill as modelled by MU–SLICKLETS*

3.5. Impact on the biota

The impact of the pollution on the biota has been calculated according to the formulae given in Section 2.7 and is summarized below.

Table 11. *Impact on the biota and corresponding losses of biomass.*

		Direct loss	*Indirect loss*	*Total loss*
Phytoplankton	11.83 %	2.12e+08 g		2.12e+08 g
Zooplankton	2.72 %	2.44e+06 g	1.06e+07 g	1.30e+07 g
Invertebrates	3.29 %	2.36e+06 g	1.72e+07 g	1.95e+07 g
Molluscs	0.61 %	4.38e+05 g	1.72e+07 g	1.76e+07 g
Demersal fishes	3.08 %	1.77e+05 g	1.60e+06 g	1.78e+06 g
Pelagic fishes	3.08 %	2.08e+05 g	1.22e+06 g	1.43e+06 g

Detailed results are also available as maps, as shown in Figure 13.

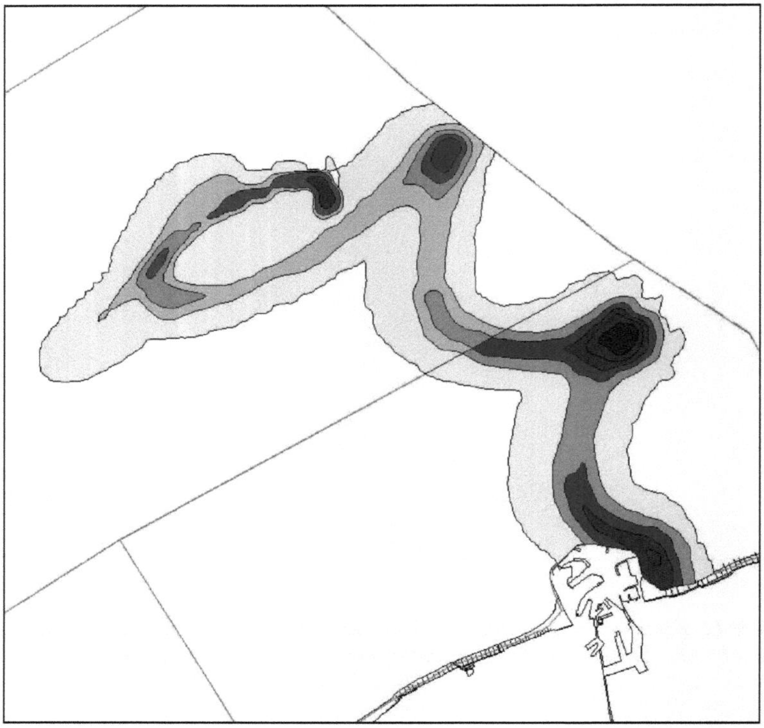

Figure 13. Spatial distribution of the benthos mortality, taking into account the ecological zonation and the distribution of oil concentration.

3.6. Impact on the seabirds

The impacted area in Zone II has a surface of 139 km² and the one in Zone Ib a surface of 84 km². Taking into account the densities of the birds in these zone in April, the highest mortality values are estimated as given on Table 12.

Table 12. *Seabird mortality per species*

Species	*Losses (individuals)*
Larus canus	27
Larus argentatus	17
Larus marinus	111
Rissa tridactyla	43
Uria aalge	81
Alca torda	17

3.7. Valuation

3.7.1. Loss of income for the fisheries

The estimate of the economic losses (€) due to the losses of biota computed under 3.5 are summarized below.

Table 13. *Conversion of biota mortality into monetary values.*

	Biomass loss (kg)	*Commercial species*	*Reference price (€)*	*Loss (€)*
Invertebrates	19500	10 %	7.44	14508.00
Demersal fishes	1780	60 %	3.72	3972.96
Total				18480.96

Note: Estimates of the economic losses based on the average fishing yield by species and taking into account the impacted zone are also computed by the model. Due to several uncertainties in the parameters available, this study case appeared to possibly be misleading and they are not reproduced here.

3.7.2. Loss of income for tourism

The maximum impact on the tourist sector occurs in the neighbourhood of Zeebrugge. There are commonly no data on tourism available for that specific part of the coast as they are unfortunately aggregated with the statistics of the city of Brugge. The model therefore returns "unknown" for the computation of the total loss (for the whole Belgian coast). Estimates of the losses for the community of Knokke alone range from 0.57 to 7.3 million euro.

3.7.3. Beach cleaning costs
Based on the volume of oil that beached on the coastline, a reference cost (euro) for the clean-up operations is computed.

Table 14. *Beach cleaning costs.*

Community name	*Sum_length*	*Sum_Beach cleaning costs*
Knokke Heist	9500	90400
Zeebrugge	8500	334000
Other	0	
TOTAL		424400

4. DISCUSSION

Models developed and/or improved in the frame of the present project have been integrated into one system.

They cover a wide range of physical, chemical and biological processes in order to describe the fate and the consequences of an accidental oil pollution at sea. Impact functions, evaluating the consequences in terms of "costs", have also been implemented.

In its present state, the system can be used to indicatively support policy options. Due to the many uncertainties and the common lack of data, we do not recommend using its results as such in application of the "polluter pays" principle, unless the underlying assumptions and approximations are clearly understood and agreed by the parties involved.

The system, however, has been implemented in a versatile manner. New developments or refinements of the existing submodels can readily be achieved and newly available data can easily be incorporated.

It is thus of the utmost importance that real cases be actively documented, in order to provide information for tuning the system.

ACKNOWLEDGEMENTS AND CREDITS

The author wishes to thank his colleagues from the other research units associated with the MARE–DASM project, who provided him with the models and the information needed to build and run the integrated system. He is also very indebted to L. VIGIN

(MUMM) for having made her skills and time available for developing the GIS post-processor.

The study project was financed by the Belgian Federal Science Policy, in the frame of the Programme 'Sustainable Management of the North Sea' (1997-2002), under contract MN/36/073 (MB/AM BA3673).

REFERENCES

- BLOKKER, P.C. (1954), *Spreading and evaporation of petroleum products on water, Rapports du 4e Congrès Portuaire International*, Antwerp, Belgium.

- DEMUTH, Cl. & SCORY, S. (1986), *ATMOS: Système de simulation du comportement des substances flottantes et volatiles déversées en mer*, Commission des Communautés Européennes, 6621/85850014. (in French).

- HEIJERICK, D. & JANSEN, C. (2002), Identificatie en kwantificatie van de verschillende factoren die bijdragen tot het degradatie van het mariene milieu, *in: MARE–DASM project, Final report.*

- KUIPERS, H. D. (1981), *SMOSS - A simulation model for oil slicks at sea,* Delft University of Technology and North Sea Directorate, Ministry of Transport and Public Works, Rijswijk, The Netherlands.

- MAC KAY, D., DI GUARDO, A., PATERSON, S., KISCI, G. & COWAN, C.E. (1996a), Assessing the fate of new and existing chemicals: a five stage process, *Environmental Toxicology and Chemistry*, 15, 1 618-1 626.

- MAC KAY, D., DI GUARDO, A., PATERSON, S. & COWAN, C.E. (1996b), Evaluating the environmental fate of a variety of types of chemicals using the EQC model, *Environmental Toxicology and Chemistry*, 15, 1 627-1 637.

- PROCTOR, R. (ED), BAART, A., BERG, P., BOON, J., DELEERSNIJDER, E., DELHEZ, E., GARREAU, P., GERRITSEN, H., JONES, J.E., DE KOK, J., LAZURE, P., LUYTEN, P., OZER, J., POLHMANN, T., RUDDICK, K., SALDEN, R., SALOMON, J.C., SKOGEN, M., TARTINVILLE, B. & VESTED, H.J. (1997), *NOMADS – North Sea Model Advection-Dispersion Study (MAS2-CT94-0105). Final Report,* POL Internal Document No. 108.

- PROCTOR, R., (ED), DAMM, P., DELHEZ, E., DUMAS, F., GERRITSEN, H., DE GOEDE, E., JONES, J.E., DE KOK, J., OZER, J., POLHMANN, T., RASCH, P., SKOGEN, M. & SORENSEN, J.T. (2002), *North Sea Model Advection-Dispersion Study 2: Assessments of model variability (MAS3-CT98-0163). Final Report,* POL Internal Document No. 144.

- REED, M., JOHANSEN, Ø., BRANDVIK, P.J., DALING, P., LEWIS, A., FIOCCO, R., MAC KAY, D. & PRENTKI, R. (1999), Oil Spill Modeling towards the Close of the 20th Century: Overview of the State of the Art, *Spill Science & Technology Bulletin*, 5 (2), 3–16.

- SCORY, S. (1982), *Etude du déplacement et de l'étalement d'une nappe d'hydrocarbures déversée en mer, sous l'effet du vent, des courants et de la houle. Travail de fin d'études,* Université de Liège (in French).

- SCORY, S. (1985), *Projet OPIA, Modèle de prévision de l'impact d'une pollution par hydrocarbures en mer,* Commission des Communautés Européennes (in French).

- SCORY, S. (1991), *The MU–Slick Model*, MUMM Report, CAMME/91/03, 28 p.

- SCORY, S. (1995), *Models used by the Belgian authorities in case of accidental spills at sea, Paper presented at the workshop on "Prediction of Short-Term Transport and Dispersion of Acidental Spills from Shipping and Off-Shore Industry,* The Hague, The Netherlands, 15-17 November 1995, 5 p.

- VAN BIERVLIET, K., BOGAERT, G., DECONINCK, M., LE ROY, D. & BOGAERT, S. (2002), Ontwikkelen van socio-economische beoordelingscriteria die het mogelijk maken de kostprijs van deze degradatie objectief te bepalen., *in: "MARE–DASM project, Final report*".

- VANDENBROELE, M., VANGHELUWE, M., JANSSEN, C., PERSOONE, G. & LE ROY, D. (1997), *Activiteitenverslag. Impulsprogramma Zeewetenschappen: Definiëring en toepassing van ecologische criteria en economische indicatoren voor de effectstudie en kostenbepaling van diverse types van verontreiniging in de Noordzee.*

- VAN DEN EYNDE, D., (1991), Refraction of swell spectra, the calculation model SPECIN, in: *MUMM Report*, 33 p. (in dutch).

- VAN DEN EYNDE, D., SCORY, S. & MALISSE, J.-P. (1995), Operational modelling of tides and waves in the North Sea on the Convex C230 at MUMM, in: *Proc. European Convex Users' Conference 1995*, Brussels, Belgium, 24-27 October 1995.

CHAPTER 10
SOCIO-ECONOMIC VALUE OF THE HUMAN ACTIVITIES IN THE MARINE ENVIRONMENT: THE BELGIAN CASE

FANNY DOUVERE

Maritime Institute, Ghent University, Belgium

Universiteitstraat 6
9000 Ghent, Belgium
Tel.: + 32 9 264 68 95/84 41, Fax: + 32 9 264 69 89
E-mail: fanny.douvere@UGent.be

1. INTRODUCTION

The Belgian Part of the North Sea (BPNS) is approximately 3,600 square km and comprises a territorial sea up to 12 nautical miles and a continental shelf, an exclusive economic zone and a fishery zone covering an area of 2,017 square km. The BPNS is the smallest part under jurisdiction of a coastal state bordering the North Sea, approximately 0.5 % of the North Sea[1]. Although small in size, the BPNS has a high degree of activity and exploitation. Sustainable development has become a central theme within the management of the marine environment. In recent years the concept has been introduced in various international agreements. Concerning the BPNS, the achievement of sustainable development has been included as a principle in the Law on the Protection of the Marine Environment[2, 3]. The application of sustainable management requires in the first place knowledge of the main human activities and their social, economic and ecological values. This

---

[1] www.mumm.ac.be
[2] Law of 20 January 1999 on the Protection of the Marine Environment, *Belgian Official Journal*, 12 March 1999.
[3] CLIQUET, A. & MAES, F., (1998), The new Belgian law on the protection of the marine environment, *Journal of International Wildlife Law and Policy*, 395-402.

F. Maes (ed.),Marine resource damage assessment, liability and compensation for environmental damage, 253–269.

chapter aims to examine the social and economic impact of the user functions in the BPNS.

2. MAIN ACTIVITIES

The principal human activities in the BPNS are shipping, dredging, disposal of dredged material, sand and gravel extraction, tourism and fisheries. Other user functions, which are less important in terms of socio-economic value, are military exercises, former sites for war munitions, telephone cables, gas pipelines and protected areas (Figure 1).

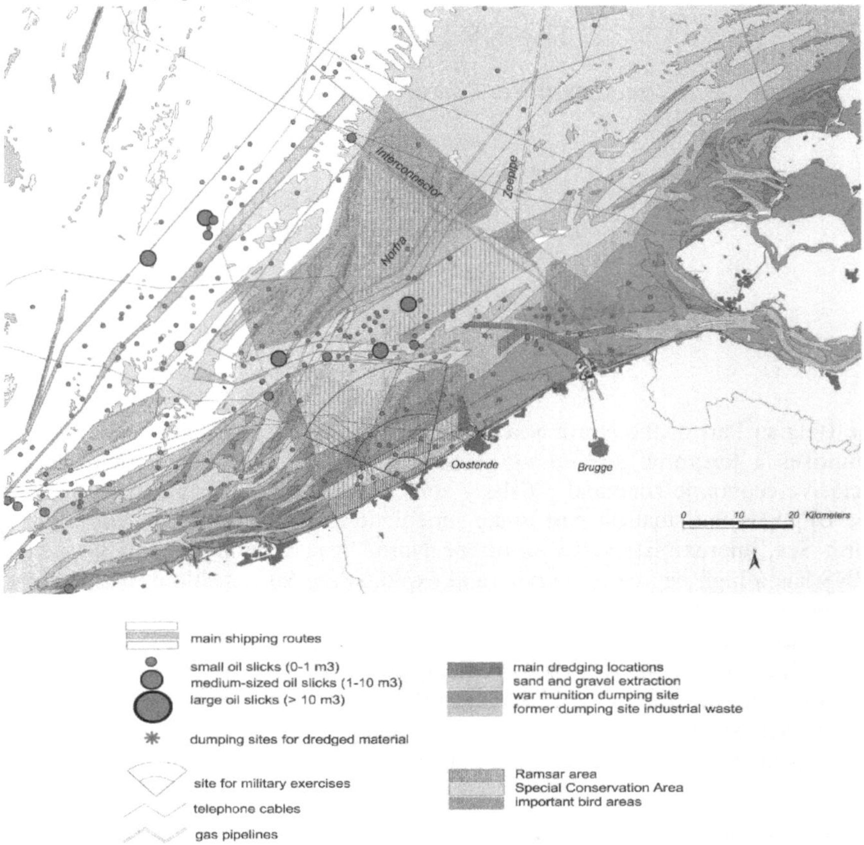

Figure 1. *Map showing the user functions in the BPNS (2002) (oil slicks 1991-1995)* [4]

---

[4] MAES, *et al.*, (2000) *Limited Atlas of the Belgian part of the North Sea*, Belgian Science Policy, 27.

Each activity with an explicit socio-economic value, *e.g.* fisheries, tourism, shipping, dredging and the extraction of sand and gravel, will be discussed in detail below. Their economic impact will be measured in terms of turnover or added value. The turnover will be measured by the parameters volume x price. The added value is related to the contribution of the user function to the Belgian economy. In case these factors are not available, the specific methodology used will be outlined in the sections concerned. The social impact is measured by the direct employment that is generated by the activity. Due to its location, the BPNS has a specific economic and social value only to the Flemish province bordering the North by the Sea, called West-Flanders. The study made use of the most recent figures available. For fisheries, tourism and shipping, 2002 has been taken as reference year. For the extraction of sand and gravel and dredging, estimates are based respectively on 2000 and 2001.

3. FISHERIES

Although fisheries contribute a small share to the national income in Belgium (2001: 0.02 % of GDP), fisheries are an important part of the human activities in the BPNS. Especially in the coastal regions, the livelihoods of people depend upon fisheries. Damage to the marine environment can directly affect their ability to earn a living and can indirectly affect business dependent on the fish catches.

In 2002, the Belgian fishing fleet consisted of 128 vessels, with a total capacity of 65,812 kW[5] and 23,992 GT[6]. Apart from fishing techniques and target species, the fleet can be divided in a small-scale fleet (vessels smaller than 221 kW) and a large-scale fleet (vessels larger than 221 kW). Half of the fleet are small-scale fishing vessels, which are primarily fishing in the BPNS[7].

The large-scale fleet is highly dependent on fishing grounds outside the BPNS. Based on the European Common Fisheries Policy (CFP), Belgium has fishing rights in other parts of the North Sea. The most important areas are under jurisdiction of the United Kingdom, Ireland, Denmark, France and the Netherlands[8]. This article concentrates on the BPNS. Therefore, only the small-scale fleet will be subject to the socio-economic assessment outlined below.

---

[5] KiloWatt.

[6] Gross Tonnage.

[7] FEDERALE OVERHEIDSDIENST MOBILITEIT EN VERVOER - MARITIEM VERVOER (2002), *Officiële lijst der Belgische vaartuigen*, 29.

[8] Council Regulation (EC) 2371/2002 on the conservation and sustainable exploitation of fisheries resources under the Common Fisheries Policy, *OJ* L 358, 2002.

A survey suggests that about 31 % of the small-scale fishing fleet is predominantly dependent for its income on the BPNS, in some cases completely[9]. This figure can be seen as an indication of the total fish catch of small-scale fishing vessels, although interviews with fishermen reveal that the fish catch is underestimated. The estimates of fish catch in the BPNS are based on data generated by use of ICES fish boxes, the borders of which are not similar to those of the BPNS. However, the data related to the ICES fish boxes are the most detailed data available today.

The Belgian fish catch is highly concentrated on cod, sole and plaice, which accounts for 60 % of the total catch[10]. Especially for small-scale fishing vessels, shrimps are an important target species. In 2001, about 16 vessels could be classified as genuine shrimp trawlers having shrimp as their main target species for at least 8 months a year. About 14 vessels could be classified as seasonal shrimp trawlers, targeting shrimp from August to November[11]. In general, fishing activities are most intensive from September to April[12].

The annual turnover strongly depends on price and volume of catch. In 2002, the total turnover of the small-scale fishing vessels was estimated at 20.6 million euro. This estimate is based on the value of landings for shrimp (total), cod, sole and plaice, of which only 31 % has been taken into account.

The small-scale fishing sector is believed to be of social rather than strong economic importance. The direct employment generated by fishery activities in the BPNS relates to the fishermen on board and their assistants on the quay. In Belgium, small-scale fishing vessels are largely managed as family businesses. Various members of the family are dependent on fish catches for their income. In general, about 3 to 5 people are working on board a small-scale fishing vessel. Interviews with fishermen and stakeholders suggest that for one person on board approximately 3 people provide assistance on land. These land based employees are mainly working in one of the 3 fishing ports with an auction, *e.g.* Zeebrugge, Ostend and Nieuwpoort. Based on these parameters, the direct employment in small-scale fisheries in the BPNS is estimated at 1,024 people (table 1).

---

9 REDERSCENTRALE (2001).

10 MINISTERIE VAN DE VLAAMSE GEMEENSCHAP, ADMINISTRATIE LANDBOUWBELEID - DIENST ZEEVISSERIJ (2002), *De Belgische Zeevisserij, Aanvoer en Besomming*, 21.

11 Polet, H., (2003) *Evaluation of by-catch in the Belgian brown shrimp (cragnon cragnon L.) fishery and of technical means to reduce discarding*, Ghent University.

12 MINISTERIE VAN DE VLAAMSE GEMEENSCHAP, ADMINISTRATIE LANDBOUWBELEID - DIENST ZEEVISSERIJ (2002), *o.c.*, 15.

Table 1. *Direct employment in small-scale fisheries, 2001*

	Direct employment
Fishermen on board	256
Assistance on land	768
Total	**1,024**

4. TOURISM

In 2002, about 20 million people visited the Belgian coast, either for a daytrip or to stay overnight. Over 80 % of the tourists are Belgian inhabitants, the others are from immediate neighbouring countries such as Germany, France and United Kingdom. The largest amount of tourists (17 million) visit the coast on a day trip basis. Day trips are especially popular during the spring and summer. More than half of the day trips take place during May and September[13]. As shown on the map below, Knokke-Heist, Blankenberge and Ostend are the most attractive cities for day-trippers (Figure 2)[14].

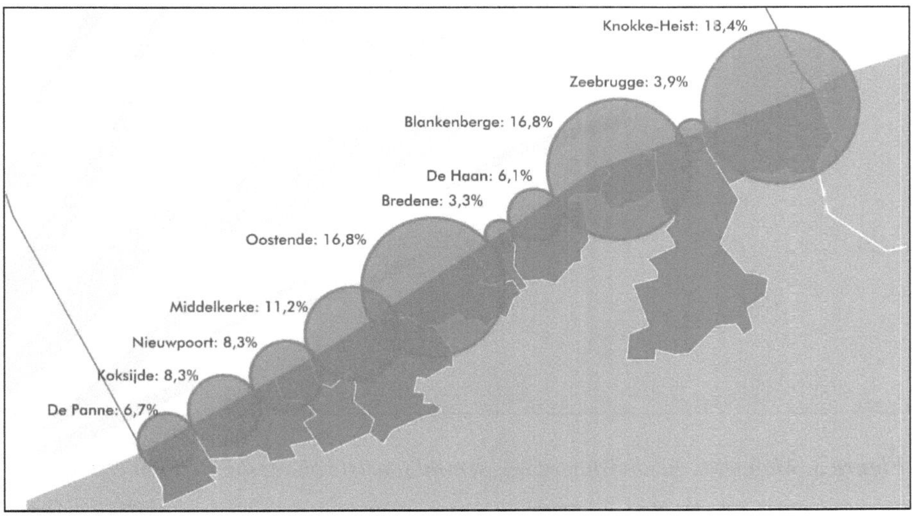

Figure 2: *Map with the division of day-trippers along the Belgian coast, 1999*

[13] WESTTOER (2002), the provincial agency for tourism and recreation in the province of West Flanders.
[14] WES, (1999).

Research suggests that day-trippers have an average spending between 10 to 20 euro per trip, per person. These figures are believed to be minimum estimates. Based on these average spendings per day, the total turnover of day-tripping to the Belgian coast is estimated between 170 and 340 million euro[15].

Next to day-trippers, ca. 3 million people visited the coast for an overnight stay in 2002. On average, tourists take residence for 6 nights, which generated near to a total of 17 million residents in commercial lodgings. Over 60 % of these tourists stay in a studio, apartment or house, rented from individual owners. Other frequented accommodations are hotels, camping sites and holiday centres or villages. Most residential tourists take their holiday at the Belgian coast between May and September. Above this, it is observed that the duration of their stay is longer in July and August compared to other periods during the year. Except for Knokke-Heist, the western part of the Belgian coast is more popular for residential tourists than the eastern part (Figure 3)[16].

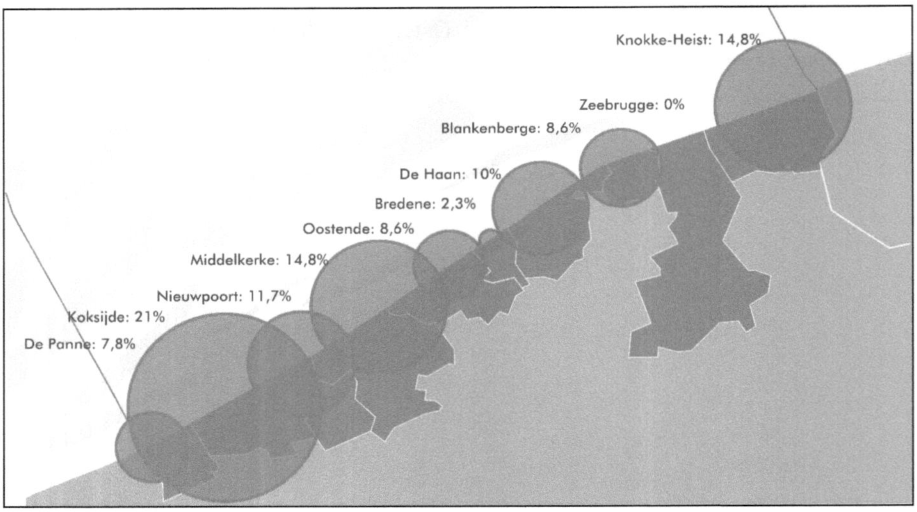

Figure 3. *Map with the division of residential tourists along the Belgian coast, 1999*

Spendings of tourists taking residence in commercial loges at the Belgian coast is about 2 to 4 times higher than day trippers. Recent surveys suggests that residential tourists are spending approximately 39.32 euro per night, per person. Based on these

[15] WESTTOER, (2002), *Strategic policy plan for tourism and recreation at the coast,* 159-160.
[16] WESTTOER, (2002), the provincial agency for tourism and recreation in the province of West Flanders.

figures, the total turnover of residential tourism in commercial lodgings is estimated as 664 million euro in 2002[17].

Apart from day-trippers and overnighters, the Belgian coast hosts about 80,000 houses or grouped residences. They are in private ownership and used as a second 'summer' residence by the owner. On average, the non-grouped summer residences are occupied by 3 to 4 people for about 66 nights, relatively equally spread throughout the year[18]. The grouped residences are occupied by 3 people for approximately 50 nights per year. It is argued that summer residences are of significant and growing economic value for the Belgian coast. Research suggests that this kind of tourism generates a total turnover of 550 million euro on a rate of 16 million nights. Approximately 40 % of these spendings are fixed costs, related to ownership and maintenance of the residence[19].

Based on the spendings above, tourists who are visiting the Belgian coast either on a day-trip basis or to stay overnight both in commercial lodgings or summer residence generated a total turnover of 1,469 million euro in 2002 (Table 2).

Table 2. *Total turnover of tourism at the Belgian coast, 2002*

	Turnover (€ million)
Day-trippers	255
Overnighters	664
Summer residences	550
Total	**1,469**

Tourism is believed to be an important provider of jobs in the coastal region, especially the accommodation and real estate sector, restaurants, pubs and the recreational facilities. The employment rate can be divided into paid and self-employment. In 2000, tourism generated total direct paid employment of ca. 8,600 jobs, which is half of the total employment in West-Flanders in the categories mentioned above. It is argued that the direct paid employment has improved significantly compared to 1992, although the overnight stays decreased by 20 % during this period (Table 3)[20].

---

17 WESTTOER, (2002), *o.c.*, 158.
18 WESTTOER, (2002), *o.c.*, 149.
19 WESTTOER, (2002), *o.c.*, 158-159.
20 WESTTOER, (2002), *o.c.*, 163.

Table 3. *Direct paid employment generated by tourism at the coast and in West-Flanders, per category, 2000*

	Tourism employment at the coast	*Tourism employment in the Province of West-Flanders*	%
Accommodation	2,562	4,235	60 %
Catering industry	3,923	10,127	39 %
Real estate sector	1,102	1,102	100 %
Recreation (incl. Tourism information services)	1,023	1,721	59 %
Total	**8,610**	**17,185**	**50 %**

Besides the paid employment, the hotel and catering industry and recreational sector accounts for about 1,869 self-employed jobs, which is a third of the self-employed business in the specified subcategories in West-Flanders (Table 4)[21].

Table 4. *Direct self- employment generated by tourism at the coast and in West-Flanders, per category, 2000*

	Touristic employment at the coast	*Touristic employment in the Province of West-Flanders*	%
Catering industry	1,605	4,579	35 %
Recreational industry	264	1,081	24 %
Total	**1,869**	**5,660**	**33 %**

5. SHIPPING

Due to the central position of Belgium, seaports and shipping are of major importance to the Belgian economy. Belgium has 4 main seaports, Antwerp, Ghent, Zeebrugge and Ostend. In 2002, these seaports transhipped over 195 million tonnes of goods. The economic and social importance of each seaport will be discussed separately[22].

[21] WESTTOER, (2002), *o.c.*, 163.

[22] VLAAMSE HAVENCOMMISSIE & SERV (2003), *Jaaroverzicht Vlaamse havens 2003,* 138.

5.1. Antwerp

Antwerp is the main seaport in Belgium. During the past decade, Antwerp has become one of the largest centres for marine chemical transport in the world. In 2002, about 16,000 ships entered the port of Antwerp, with a total gross tonnage of 218 million. The total traffic of goods rose to 143 million tonnes. Almost half of the traffic concerns containers and dry mass goods (43.8 %). Goods transhipped to Antwerp are mainly coming from Europe (38 %). The dominant destinations are Asia (33 %) and Europe (30 %).

The port of Antwerp generated a total added value of 7 billion euro in 2002. The non-maritime cluster[23] is responsible for 78 % of the total added value. More than half of it is industrial, which includes petroleum, chemical, metal, car manufacturing and electricity industry. The maritime cluster[24] has a share of 22 % in the total added value (Table 5).

Table 5. *The total added value of the seaport of Antwerp, by category 2002*

	Added value (million euro)	*%*
Wholesale trade	772.8	11 %
Industry	4,036.2	58 %
Logistics	435.9	6 %
Transport	209	3 %
Total non-maritime cluster	**5,453.9**	**78 %**
Maritime cluster	1,559	22 %
Total	**7,012.9**	**100 %**

In 2002, Antwerp provided over 60,000 jobs. The major suppliers are the industrial part of the non-maritime cluster (45 %), followed by the maritime cluster (35 %). Wholesale trade, logistics and transport are of second importance in terms of creation of employment (Table 6).

[23] The non-maritime cluster contains companies of industry, trade, transport and logistics which are economically not directly related to the seaport activities, but the activities of which are partly dependent on the geographical proximity of the seaport.

[24] The maritime cluster refers to all companies directly linked with the activities in the seaport of Antwerp and with activities that are essential for the functioning of the seaport.

Table 6. *The total employment of the seaport of Antwerp, by category, 2002*

	Total employment	*%*
Wholesale trade	2,654	4 %
Industry	27,707	45 %
Logistics	5,597	9 %
Transport	3,449	7 %
Total non-maritime cluster	**39,407**	**65 %**
Maritime cluster	21,156	35 %
Total	**60,563**	**100 %**

5.2. Ghent

In 2002, approximately 2,800 ships entered the seaport of Ghent, with a total gross tonnage of 28 million. More than 23 million tonnes of goods were transhipped in the port of Ghent, of which over 80 % has been unloaded. About 66 % of the transhipments concern dry mass goods. Goods transported to Ghent come mainly from South America (33 %) and Europe (29 %). Europe is the main destination (88 %) of the goods shipped.

The port of Ghent generated a total added value of 2.9 billion euro in 2002. The contribution of the non-maritime cluster, the industry in particular, is 93 % (Table 7).

Table 7. *The total added value of the seaport of Ghent, by category, 2002*

	Added value (million euro)	*%*
Wholesale trade	592.5	20 %
Industry	1,959.4	67 %
Logistics	112.0	4 %
Transport	64.9	2 %
Total non-maritime cluster	**2,728.8**	**93 %**
Maritime cluster	214.1	7 %
Total	**2,942.9**	**100 %**

In 2002, the seaport of Ghent provided 28,500 jobs. In line with the added value, 92 % is generated by the non-maritime cluster, of which 73 % comes from the industry (Table 8).

Table 8. *The total employment of the seaport of Ghent, by category, 2002*

	Employment	*Share*
Wholesale trade	2,851	10 %
Industry	20,923	73 %
Logistics	1,401	5 %
Transport	1,014	4 %
Total non-maritime cluster	**26,189**	**92 %**
Maritime cluster	2,312	8 %
Total	**28,501**	**100 %**

5.3. Zeebrugge

Zeebrugge is the third largest seaport of Belgium (in terms of added value). In 2002, about 10,000 ships entered the seaport, with a total gross tonnage of 89 million. About 33 million tonnes goods have been transhipped in the port of Zeebrugge. Europe, and the United Kingdom in particular, are the main origins (63 %) and destinations (77 %) of goods transhipped in Zeebrugge. Besides goods, passengers are also transhipped in Zeebrugge. About 670,000 people were transported in 2002, also primarily from and to the United Kingdom.

In 2002, the port of Zeebrugge generated a total added value of 687 million euro. The non-maritime cluster (61 %), and the industry in particular, are the largest contributor to the economic value (Table 9).

Table 9. *The total added value of the seaport of Zeebrugge, by category, 2002*

	Added value (million euro)	*%*
Wholesale trade	41.2	6 %
Industry	272.2	40 %
Logistics	45.2	7 %
Transport	57.3	8 %
Total non-maritime cluster	**415.9**	**61 %**
Maritime cluster	271.1	39 %
Total	**687.0**	**100 %**

In 2002, the seaport of Zeebrugge generated close to 10,000 jobs. In comparison with other seaports, employment is more equally divided amongst the two clusters.

The non-maritime cluster provides 53 % of the jobs, the maritime cluster 46 %. Wholesale trade, logistics and transport are of secondary importance (Table 10).

Table 10. *The total employment of the seaport of Zeebrugge, by category, 2002*

	Total employment	*%*
Wholesale trade	677	7 %
Industry	2,878	29 %
Logistics	755	8 %
Transport	942	10 %
Total non-maritime cluster	**5,252**	**54 %**
Maritime cluster	4,533	46 %
Total	**9,783**	**100 %**

5.4. Ostend

Ostend is the smallest Belgian seaport. In 2002, about 4,200 ships entered the port of Ostend, with a total gross tonnage of 33 million. Over 6 million tons of goods were transhipped, of which roll-on/roll-off is the most important maritime transport (70 %). About 18 % of the transport concerns the supply of sand and gravel by offshore extraction. Almost the whole trade is intra-European. Next to Zeebrugge, Ostend is also important for the transhipment of passengers. About 149,000 people were transported in 2002, mostly from and to the United Kingdom.

In 2002, the port of Ostend generated a total added value of 324 million euro. The non-maritime cluster generated 82 % of the total economic value, in which industry has the largest share (Table 11).

Table 11. *The total added value of the seaport of Ostend, by category, 2002*

	Total added value (million euro)	*%*
Wholesale trade	21.7	7 %
Industry	168.5	52 %
Logistics	53.2	16 %
Transport	22.3	7 %
Total non-maritime cluster	**265.7**	**82 %**
Maritime cluster	58.3	18 %
Total	**324**	**100 %**

Over 4,000 people were employed in the port of Ostend in 2002. The non-maritime cluster generated almost 80 % of this employment. The wholesale trade, logistics and transport are also in Ostend of second importance (Table 12).

Table 12. *The total employment of the seaport of Ostend, by category, 2002*

	Total employment	*%*
Wholesale trade	367	9 %
Industry	1,758	43 %
Logistics	814	20 %
Transport	296	7 %
Total non-maritime cluster	**3,235**	**79 %**
Maritime cluster	860	21 %
Total	**4,095**	**100 %**

5.5. Conclusions

Antwerp, Ghent, Zeebrugge and Ostend transhipped over 195 million tonnes of goods and transported about 827,000 passengers[25]. The seaports created an overall added value of 11 billion euro. The share in the total added value of Antwerp, Ghent, Zeebrugge and Ostend is respectively 64 %, 27 %, 6 % and 3 %. More than 80 % of the total added value of the Belgian seaports is generated from the non-maritime cluster, in particular the industrial sector (Table 13).

In 2002, the 4 seaports generated a total direct employment of approximately 103,000 people. As illustrated above, more than half of the employees are working in the industry. In general, the maritime cluster, the wholesale trade, the transport and the logistics are of secondary importance for job creation (Table 13).

[25] This figure contains the passengers of Antwerp and Ghent, which are not mentioned in the detailed survey above.

Table 13. *Share of the Belgian seaports in the total direct employment and added value, 2002*

	Direct Employment	*%*	*Added Value to Belgian economy (€ million)*	*%*
Antwerp	60,563	59 %	7,012.9	64 %
Zeebrugge	9,783	9 %	687	6 %
Ghent	28,501	28 %	2,942.9	27 %
Ostend	4,095	4 %	324	3 %
Total	**102,942**	**100 %**	**10,966.8**	**100 %**

6. DREDGING ACTIVITIES

Due to the high frequency of the shipping to Belgian seaports, maintenance dredging is an important activity in the BPNS. Dredging activities can be divided into land reclamation, deepening and investment dredging and maintenance dredging. Dredging activities in the BPNS are subcontracted by the Flemish government to a Temporary Corporation *'Noordzee en kust'*, which is composed of a consortium of private companies.

In 2001, 10.2 million m^3 has been dredged (maintenance and land reclamation) in the BPNS. Over 60 % was carried out to maintain the navigation routes to the Belgian seaports[26]. Most of the dredged material is dumped on specified locations in the BPNS.

The turnover of dredging activities in the BPNS is estimated by making use of the budget spent by the Flemish government. Based on this figure, in 2001, dredging activities generated a global turnover of 56.8 million euro. About 85 % of the turnover is generated by maintenance dredging.

Research suggests an employment rate in the dredging sector between 240 and 560 people in 2001. Due to the specific structure of the companies involved, the employment related to dredging activities in the BPNS can only be roughly estimated. To estimate job creation in the dredging sector, one has to depart from the budget spent by the Flemish government for dredging activities. Further interviews with stakeholders in the sector reveals that the yearly budget for dredging activities in the BPNS is composed as follows:

[26] MAES, F., DOUVERE, F., SCHRIJVERS, J., (2002), Ontwikkelen en evalueren van maatregelen die de overheid dient te nemen om een geïntegreerd beheer en duurzaam gebruik van de Noordzee te bevorderen, in: *Beoordeling van mariene degradatie in de Noordzee en voorstellen voor een duurzaam beheer*, Universiteit Ghent, 502.

$$P=Pox(0,37\,S + 0,13\,G + 0,18\,I + 0,32)$$

(S) = labour cost
(G) = fuel
(I) = index

In 2001, total labour costs were 12.5 million euro, an average hourly wage (12.6 euro net) and an average amount of working hours per year reveals a rough indication of 560 employees working on dredging activities in the BPNS.

Based on interviews with stakeholders of the companies involved, the employment rate was estimated at 240 jobs. About 40 % of the employees had a higher education. Over 65 % of them were working on board dredging vessels. Other jobs were counted in administrative, maintenance and quayside employment[27].

7. SAND AND GRAVEL EXTRACTION

The extraction of sand and gravel on the BPNS takes place in two concession zones (Figure 4). Sand and gravel extraction is only permitted in these zones for companies that hold a concession from the government. In 2001, 16 companies received permission to extract about 1.7 million m^3 sand and gravel. The largest amount of sand and gravel is used for construction works. Other uses are beach suppletion and the restraint of coastal erosion.

In 2000, the total turnover from offshore sand and gravel extraction was estimated at 125.5 million euro. This figure is based on the total turnover of companies holding a concession[28].

In this period, sand and gravel activities generated about 22,000 jobs, however not all involved offshore sand and gravel extraction. Due to the structure and organisation of the companies, no distinction between direct and indirect employment has been made. Most of the concession holders are involved in related activities, in and outside Belgium. The assessment of the employment rate for the sand and gravel extraction has been counted by the number of jobs involved in[29]:

(i) the extraction of sand;
(ii) the extraction of gravel;
(iii) the extraction of clay and kaolin;
(iv) the process industry of concrete (for the construction industry).

---

27 MAES, F., DOUVERE, F., SCHRIJVERS, J., (2002), *o.c.*, 512.
28 NATIONALE BANK VAN BELGIË - DIENST BALANSCENTRALE (2002).
29 MAES, F., DOUVERE, F., SCHRIJVERS, J., (2002), *o.c.*, 502.

Interviews with stakeholders suggests that the indirect employment (processing industry) is responsible for at least half of the jobs indicated above.

8. CONCLUSIONS

The total economic value of the main human activities in the BPNS is estimated at 12.6 billion euro. As expected, the shipping sector has the largest contribution to this global economic value (86.8 %). Tourism is responsible for about 12 % of the added value. Especially the accommodation and real estate sector, restaurants and pubs and the recreational facilities depend very much on coastal tourism. Although highly influenced by related (indirect) activities, almost 10 % of the economic value is generated by the sand and gravel sector. Only 0.5 % of the added economic value is generated by dredging activities. Fishery activities are small in terms of economic value (2001: 0.2 % of GDP) (Table 14).

The key user functions on the BPNS provided a total direct employment of approximately 137,000 jobs. Three quarters of all employment is generated by the 4 seaports, primarily in the industry sector of the non-maritime cluster. The offshore extraction of sand and gravel is responsible for about 16 % of jobs, although research suggests that half of these employees are working for the indirect processing industry (which is not included in the estimates of the other user functions). Although small in numbers (0.8 %), fisheries are of high social importance for the coastal region. Dredging activities have the smallest share in the total coastal employment (Table 14).

Table 14. *The global social and economic impact of the user functions in the BPNS, 2002*

	Total added value or turnover (€ million)	*%*	*Total employment (jobs)*	*%*
Fisheries	20.6	0.2 %	1,024	0.8 %
Tourism	1,469.0	11.6 %	10,479	7.6 %
Shipping	10,966.8	86.8 %	102,942	75.2 %
Dredging*	56.8	0.5 %	400	0.3 %
Sand and gravel**	125.5**	**0.9 %**	22,000***	16.1 %
Total	**12,638.7**	**100 %**	**136,845**	**100 %**

* Figures for 2001
** Figures for 2000
** This figure includes the indirect added value of the process industry of concrete (construction sector)
*** This figure includes the indirect employment of the process industry of concrete (construction sector)

REFERENCES

- Council Regulation (EC) 2371/2002 on the conservation and sustainable exploitation of fisheries resources under the Common Fisheries Policy, *OJ* L 358, 2002.
- Law of 20 January 1999 on the Protection of the Marine Environment, *Belgian Official Journal* 12 March 1999.

- FEDERALE OVERHEIDSDIENST MOBILITEIT EN VERVOER - MARITIEM VERVOER, *Officiële lijst der Belgische vaartuigen.*
- MAES, F. *et al.*, (2000), *Limited Atlas of the Belgian part of the North Sea*, Belgian Science Policy, 31 p.
- MAES, F., DOUVERE, F. & SCHRIJVERS, J., (2002), Ontwikkelen en evalueren van maatregelen die de overheid dient te nemen om een geïntegreerd beheer en duurzaam gebruik van de Noordzee te bevorderen, in: *Beoordeling van mariene degradatie in de Noordzee en voorstellen voor een duurzaam beheer*, Universiteit Ghent.
- MINISTERIE VAN DE VLAAMSE GEMEENSCHAP - ADMINISTRATIE LANDBOUWBELEID - DIENST ZEEVISSERIJ (2002), *De Belgische Zeevisserij, Aanvoer en Besomming.*
- NATIONALE BANK VAN BELGIË - DIENST BALANSCENTRALE (2002).
- POLET, H. (2003), *Evaluation fo by-catch in the Belgian brown shrimp (cragnon cragnon L.) fishery and of technical means to reduce discarding*, Ghent University.
- REDERSCENTRALE (2001).
- VLAAMSE HAVENCOMMISSIE & SERV (2003), *Jaaroverzicht Vlaamse havens 2003.*
- WESTTOER (2002), *Strategic policy plan for tourism and recreation at the coast.*

INDEX

MIX
Papier aus verantwortungsvollen Quellen
Paper from responsible sources
FSC® C105338

FSC
www.fsc.org

If you have any concerns about our products,
you can contact us on
ProductSafety@springernature.com

In case Publisher is established outside the EU,
the EU authorized representative is:
Springer Nature Customer Service Center GmbH
Europaplatz 3, 69115 Heidelberg, Germany

Printed by Libri Plureos GmbH
in Hamburg, Germany